POCKET-BOOK

FOR

RAILROAD AND CIVIL ENGINEERS.

EARTHWORKS.

Excavating.

FIG.77.

Longitudinal Section of Fig.78

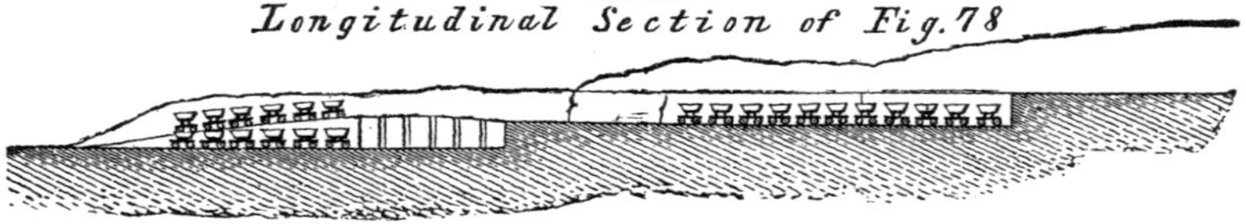

FIG.78.

Plan of Fig. 77.

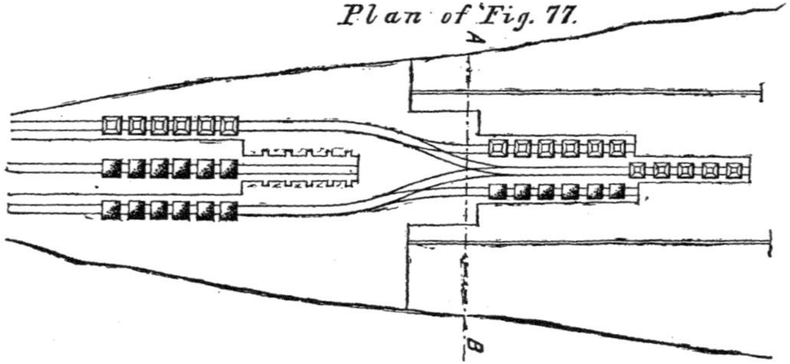

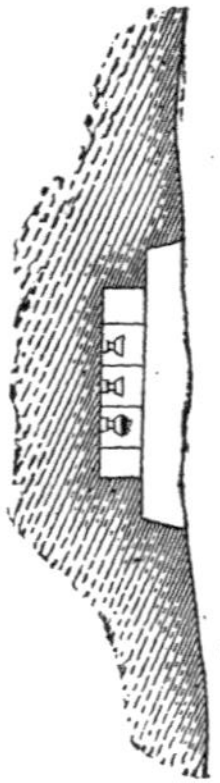

FIG.79

Section on A.B. of Fig.78.

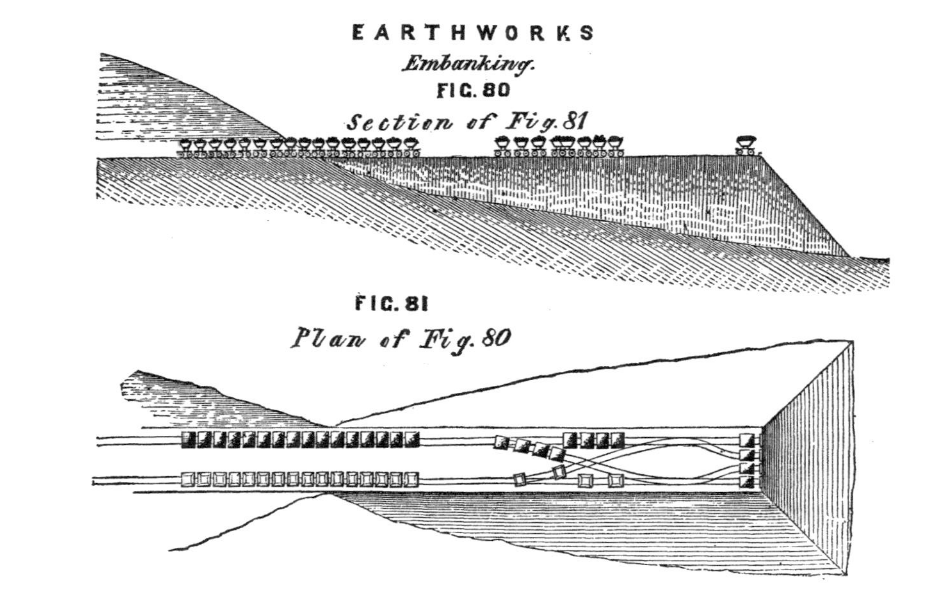

EARTHWORKS

Embanking.

FIG. 80

Section of Fig. 81

FIG. 81

Plan of Fig. 80

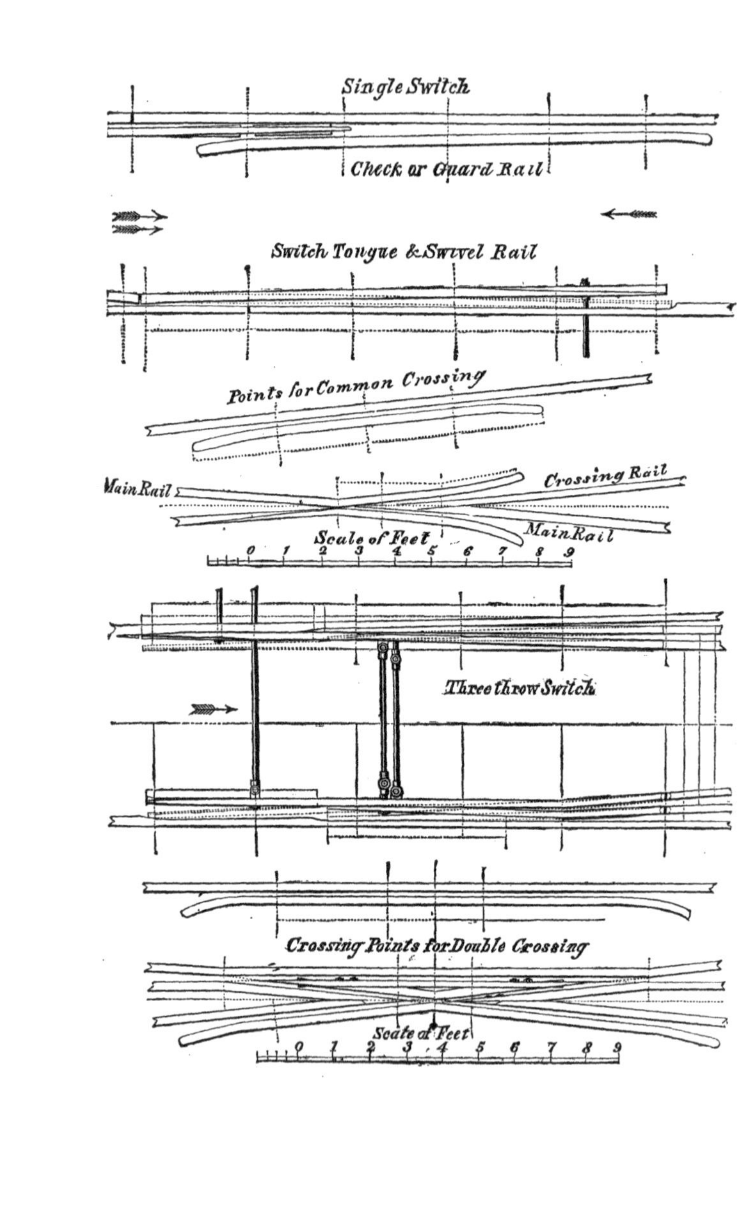
Single Switch
Check or Guard Rail
Switch Tongue & Swivel Rail
Points for Common Crossing
Main Rail
Crossing Rail
Main Rail
Scale of Feet
0 1 2 3 4 5 6 7 8 9
Three throw Switch
Crossing Points for Double Crossing
Scale of Feet
0 1 2 3 4 5 6 7 8 9

See page 92.

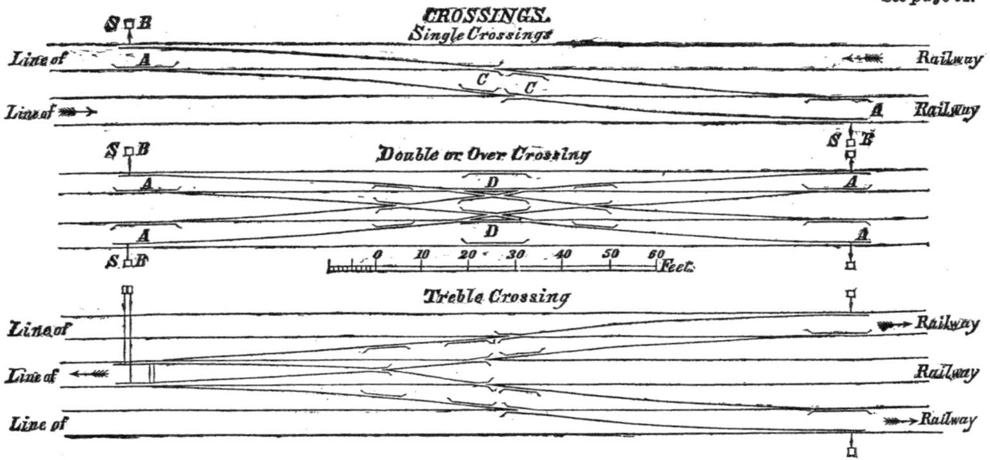

CROSSINGS.

Single Crossings

S B
Line of A Railway
C C
Line of A Railway
S B

Double or Over Crossing

S B
A D A
A D A
S B

0 10 20 30 40 50 60 Feet.

Treble Crossing

Line of Railway
Line of Railway
Line of Railway

POCKET-BOOK

FOR

RAILROAD AND CIVIL ENGINEERS.

CONTAINING

NEW, EXACT, AND CONCISE METHODS FOR LAYING OUT RAILROAD CURVES, SWITCHES, FROG ANGLES, AND CROSSINGS; THE STAKING OUT OF WORK, LEVELLING; THE CALCULATION OF CUTTINGS AND EMBANKMENTS, EARTHWORK, ETC.

BY OLIVER BYRNE,

CIVIL, MILITARY AND MECHANICAL ENGINEER.

Author of "Appletons' Dictionary of Machines, Mechanics, Engine Work and Engineering," "The Practical Metal Workers' Assistant," "The American Engineers' Assistant," "The Pocket Companion for Machinists, Mechanics and Engineers."
Also the Author of Numerous other Mechanical and Mathematical Works, Published both in Europe and America.

PHILADELPHIA:
HENRY CAREY BAIRD,
INDUSTRIAL PUBLISHER,
No. 406 WALNUT STREET.
1864.

PREFACE.

THIS work could only be produced after long experience and much practice in Railroad and Civil Engineering. I commenced my practical operations with the construction of the first railroad of our modern system, and after passing through the different practical gradations, became the chief acting and consulted Engineer of many extensive lines of road. All writers on this subject, and compilers of pocket-books for railroad practice, without exception, fill their works with tables too contracted for use, formulas and rules too complex for practical men, or empiricisms that give results not sufficiently exact for practical purposes. It is well known that the simplest calculation cannot be made on the ground, with any chance of accuracy, without retiring to a tent or a house ; the most the Engineer can do is to store his field-book with topographical or instrumental notes, or with simple angular and linear measurements ; hence all helps to perform extensive calculations on the ground are useless.

Few experienced accomplished Engineers ever employ a table of logarithms found in a Pocket-book to perform a series of extensive calculations for engineering purposes, unless compelled by great necessity.

On the ground, however, many things must be remembered, so many, that few can retain them in the memory. This Pocket-book contains all that I found necessary, in my extensive practice, to be remembered, and may be found of as much use to other Engineers.

To refresh the memory, I have in many cases resorted to plain arithmetical examples instead of algebraic formulas, although the latter method is more universal. During my great and long experience of thirty years, applying the sciences to the arts, I have found this course the most effectual.

Many of the plans and rules introduced in this work are peculiarly my own, and appear for the first time in this work ; they are not empirical, and will be found great economizers of time and labor ; in this place I will only allude to

1. The laying out of railroad curves by ordinate tables of whole numbers.

2. How to drive side-stakes exactly, without trial and error

3. When the cross-sections of cuttings and embankments are irregular, a rule is given to find exactly, the height of equivalent level cross sectional areas.

4. A general Earth-work table without supposing the side slopes to meet under the centre of the road.

5. The proper coning of wheels of railroad cars, and the true rise of the outer rail on curves.

6. Calculations of cuttings and embankments :—No other pocket-book contains general tables for this purpose ; and no work contains tables so general, accurate, and extensive. To perform these calculations has been the greatest mental labor in railroad practice. The present work has been prepared with scrupulous care and accuracy, for which I trust my reputation as an Author, Engineer, and Mathematician will be a guarantee.

OLIVER BYRNE.

CONTENTS.

POCKET-BOOK

FOR

RAILROAD AND CIVIL ENGINEERS.

PROPOSITION I.

To find the Radius of a Circular Railroad Curve, the straight portions of the Road being given.

Curved rulers, segments of circles, made of hard wood, strong pasteboard or of thin sheet metal, with their radii in inches marked on them, are employed to determinethe most favorable ground for a railway, when the survey is made and laid down. By these shapes or curves, sold by every mathematical instrument-maker, or easily cut out of veneer and rounded by sand paper, the radii of railway curves are commonly found in practice; but as a circular arc of great radius apparently coincides with its tangent for a considerable distance, it is difficult, in this way, to determine the exact starting points of curves; besides, if the survey or plan be not strictly accurate, as is too often the case, the positions of the straight parts of the road, or of the tangents, cannot be said to be given; hence this plan of mechanically laying down railroads only roughly determines the radii and starting points of curves, so that the railroad may, by more accurate calculations and observations, fall, in its progress, on the most suitable ground.

Fig. 1.

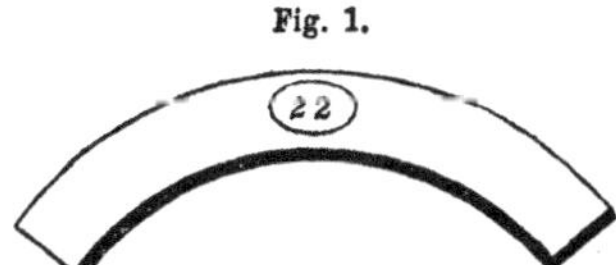

Fig. 1 represents a curved ruler of 22 inches radius. If the map or plan to which this ruler is applied, be laid down to a scale of 6 chains to the inch, it will represent a circular arc of $22 \times 6 =$ 132 chains radius. If the scale of the plan be 4 chains to an inch it will represent an arc whose radius is 88 chains $= 4 \times 22$. These curves are sold in sets, beginning with a radius of $2\frac{1}{2}$ inches, and terminating with one of 160 inches or upwards. No mistake can be made, as the radius in inches is marked on each circular curved ruler.

Fig. 2.

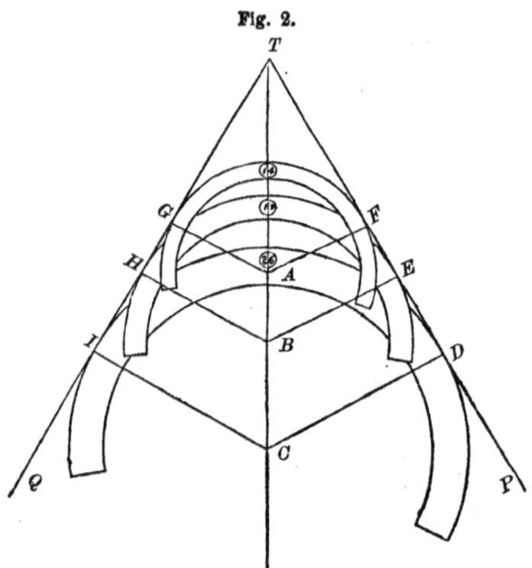

Let QI and PD be two straight portions of a railroad, the positions of which are given by an accurate plan; and which when prolonged meet at T. Apply several of the series of curved rulers to touch the line QT, without cutting it, at the points G, H, I, and also to touch PT, in the same way, at the points D, E, F. Then on a careful examination, the curve is adopted, which is least obstructed by hills, lakes, rivers, buildings, &c. The radius is determined by multiplying the number on the curved ruler by the number of chains an inch on the scale of the plan.

EXAMPLES.

Suppose the scale of the plan to be 5 chains to the inch, then he radius of the curve ID will be

26 × 5 = 130 chains = CD.
HE, 19 × 5 = 95 chains = BE.
GF, 14 × 5 = 70 chains = AF.

The curved rulers are marked 14, 19, 26, that is, they belong to circles of 14, 19 and 26 inches radii. Some engineers employ a chain 66 feet long, containing 100 links, of 7·92 inches each; this chain is called Gunter's chain. Other engineers employ a chain 100 feet long; but for laying out railway curves a 50 foot chain is more convenient than either of the other two.

The radius of a railroad curve may be found geometrically thus: Produce QI, PD, till they meet at T; bisect the angle QTP, by the line TC, and assuming I the starting point of the curve, draw IC perpendicular to QT, meeting TC in C. Then C is the centre and CI = CD the required radius of the railroad curve ID. Or the construction may be thus made: Suppose H the starting point; make HT = TE; draw the perpendiculars HB, EB, meeting in B; then BH or BE is the radius of the curve. When any other starting point, as G, is assumed, the radius may be found in the same manner. It is clear that this problem admits of an infinite number of solutions, but the curve adapted, is, as before observed, that which falls on ground presenting the fewest natural or artificial obstructions, provided the practical limit be not passed. Supposing the map to be inaccurate, when the starting point P, Fig. 3, is given, the radius of the curve may be found by taking dimensions on the ground.

Fig. 3.

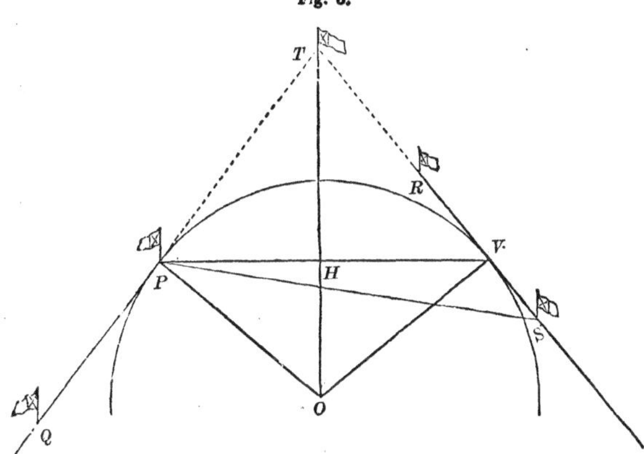

Range the tangents PQ, RS, till they meet at T; measure PT, then measure on TS till TV = TP; then measure VP.

$$\text{The radius PO} = \tfrac{1}{2}\,(PV \times PT) \div \sqrt{PT^2 - \tfrac{1}{4}\,PV^2}.$$

Before showing how this expression is found we shall give a practical example.

EXAMPLE.

Let PT = 100 chains, and PV = 120 chains, required the radius PO.

$$PO = \tfrac{1}{2}\,(120 \times 100) \div \sqrt{(100)^2 - \tfrac{1}{4}\,(120)^2} = 6000 \div 80 = 75 \text{ chains}$$

THEORY.

In the triangle PTH, we have,

$$TH = \sqrt{PT^2 - PH^2} = \sqrt{PT^2 - \tfrac{1}{4} PH^2}.$$

The triangles PHT, and PHO are similar.

Therefore, TH : PT : : PH : PO, that is,

$$\sqrt{PT^2 - \tfrac{1}{4} PH^2} : PT :: \tfrac{1}{2} PV : PO.$$

$$\therefore PO = \tfrac{1}{2} (PT \times PV) \div \sqrt{PT^2 - \tfrac{1}{4} PH^2}$$

which demonstrates the truth of the rule applied.

EXAMPLE.

Having determined by a well constructed plan, that a circular curve of 109 chains radius will suit to connect the straight portions PQ, SV, of a railroad; what is the length of the tangent PT or TV, supposing PV to measure 182 chains?

PV must be so measured that the angle TPV = angle TVP; this is easily established, by selecting any point S, and taking by the Theodolite the angles

RSP and TPS;

half the sum of these angles will give the angle TPV or TVP.

$$OH = \sqrt{OP^2 - PH^2} =$$
$$\sqrt{109^2 - 91^2} = 60$$

The triangles POH and TOP being similar,

$$HO : HP :: PO : PT$$
$$60 : 91 :: 109 : \frac{109 \times 91}{60} = 165{\cdot}31$$

165 chains 31 links.

PROPOSITION II.

To find the Radius of a Circular Railroad Curve, when the plan or map is inaccurate, and the Tangents AB, PC, cannot be prolonged to meet at T on account of Obstructions, the starting point B of the Curve being given. Fig. 4.

On the most convenient ground, measure a line BP, to meet CP in P, take the angles TBP, TPB; the sum of which taken from 180 degrees gives the angle BTP. Then by trigonometry we have

Sin. BTP : BP : : Sin. TPB : BT.

Fig. 4.

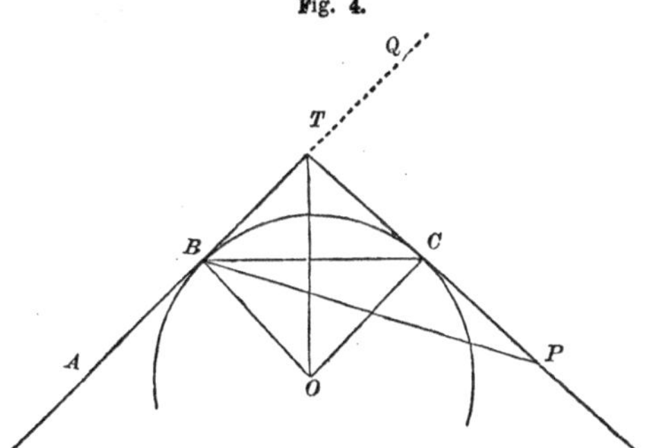

Again in the triangle BTO, right angled at B, we have given BT, and the angle BTO = half BTP, to find the radius OB; angle BOT = 90° — BTO.

Sin. BOT : BT : : Sin. BTO : OB.

Or cot. BTO : radius : : BT : BO.

From either of these proportions BO may be determined.

PRACTICAL EXAMPLE.

Suppose BP to measure 225 chains, and that the angle TBP = 62° 44′, and BPT = 25° 16′, what is the length of the tangent TB, and radius BO?

	° ′		° ′	
	62 44		180 00	
	25 16		98 00	
angle QTC =	98 00	2)	82 00	= BTC.
			41 00	= BTO.

It often happens in practical operations, that the use of natural sines and tangents are more convenient and satisfactory than the application of Log. sines and Log. tangents; as the latter are merely the logarithms of the former with their indices increased by 10, proportion is worked out by addition and subtraction when logarithms are used, but by multiplication and division when natural sines, cosines, tangents, &c., are employed.

Nat. sin. BTP = sin. 82° = .99027
Nat. sin. TPB = sin. 25° 16′ = .42683

·99027 : 225. : : ·42683. : 96 chains 98 links.
225.

213415
85366
85366

·99027 96·03675(96·98 = BT.
891243

691245
594162

970830
891243

795870
792216

Nat. sin. BTO = sin. 41° = ·65606
90° — 41° = 49° = BOT; sin 49° = ·75471
Natural cotangent BTO = cot. 41° = 1·15037

·75471 : 96·98 : : ·65606 : 84 chains 30 links.
96·98

524848
590454
393636
590454

63·6246988

·75471)63·6246988(84·30 = BO.
603768

324789
301884

229058
226413

26458

OB, the radius may be found thus—

1·15037 : 1· : : 96·98 : 84·30 = BO.

1·15037)96·98000(84·30 chains.
920296

495040
460148

348920
345111

CALCULATION BY LOGARITHMS.

	Logarithms.	
As sin. 82°	9·9957523	Sub.
: 225 chains	2·3521825	Add.
: : sin. 25° 16′	9·6302568	
	11·9824393	
: 96·98 chains	1·9866870	

BY ADDITION ONLY.

Arithmetical complement of sin. 82° =

9·9957523	= 0·0042477
Log. 225	= 2·3521825
Log. sin. 25° 16′	= 9·6302568
Log. 96·98	= 1·9866870
As cot. 41°	10·0608369
: radius	10·0000000
: : 96·98 chains	1·9866870
	11.9866870
: 84·30	1·9258501

PROPOSITION III.

To define the principal properties of the circle and its tangents and chords, that relate to railroad engineering.

FIRST.

The radius drawn to the point of contact is perpendicular to the tangent : OB is perpendicular to BQ. (Fig 4.)

SECOND.

The angle of intersection, QTC, Fig. 4, of any two tangents, as AB, CP, is equal to the angle COB at the centre.

THIRD.

Equal chords subtend equal angles at the circumference of a circle, as well as at the centre. Angle 1,B,2, = 2,B,3, = 3,B,4, = 4,B,5, &c. ; B,O,1, = 1,O,2, = 2,O,3, = &c. ; when B,1, = 1,2, = 2,3, = 3,4, = 4,5, = &c.

Fig. 5.

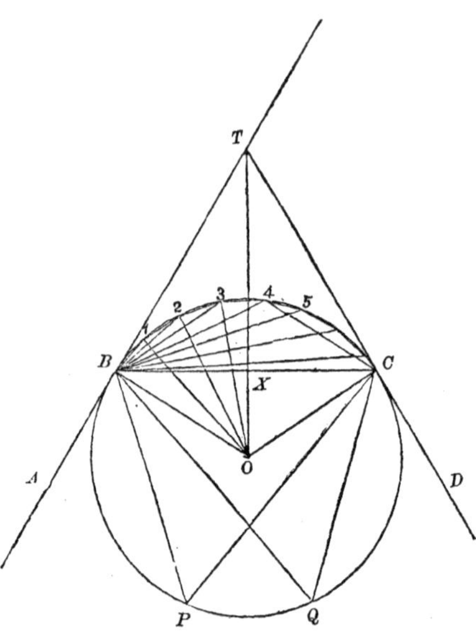

FOURTH.

Any chord BC, drawn from a point of contact B, cuts off a segment of a circle CQPB; and any angle in this segment as CQB, or CPB, is equal to the angle TBC, between the chord and the tangent. Also angle CBA = B,4,C, or any other angle contained in the segment B,1,2 3, C. (Fig. 5.)

FIFTH.

The tangents TB, T,C, drawn from any point T, are equal and the angle TBC = TCB.

SIXTH.

The angle at the centre of a circle is double the angle at the circumference subtended by the same chord.

Angle BOC = double BPC or BQC;
Angle 1,O,2 = double 1,B,2;
Angle 2,O,3 = double 2,B,3.

SEVENTH.

An angle TBC, between a tangent and a chord, is equal to BOT, half the central angle subtended by the same chord. The right angled triangles BOX, BTX, and BTO are similar. The sum of the angles BOC, BTC, equal 180 degrees.

EIGHT.

If two straight lines IF, CG, cross one another within a circle, the rectangle under the segments CK and KG is equal to the rectangle under the segments IK, KF. (Fig. 6.) If BH cross IF at right angles, IJ = JF, and BJ × JH = JF².

Fig. 6.

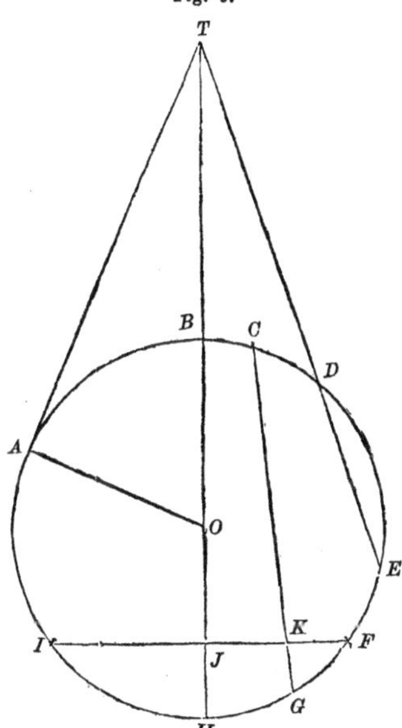

NINTH.

If from a point T outside a circle, two straight lines TH and TE be drawn to cut it and another, TA, drawn to touch it, then TH × TB = TA² = TE × TD. (Fig. 6.)

Fig. 7.

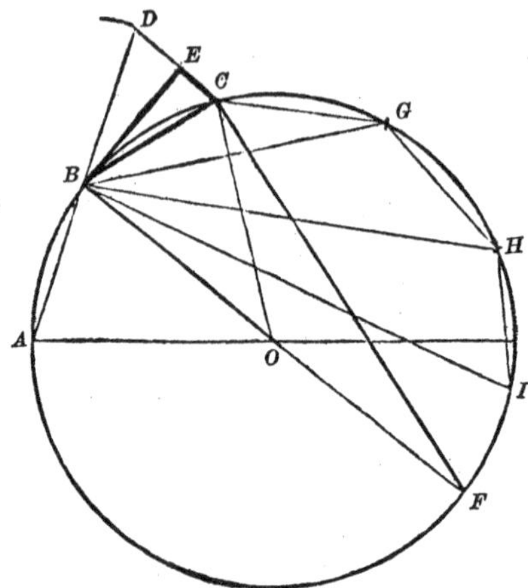

TENTH.

Take any two consecutive equal chords AB, BC, in a circle, and produce one of them as AB, till BD, the part produced is equal to AB; then join DC and bisect it in E, draw BE; and BE is a tangent to the circle at B.

The foregoing nine properties are established in the first six books of Euclid, and require no proof here, but the tenth property not being established in the elements of geometry, I shall add a proof. (Fig. 7.)

Angle ABO = OBC,
Angle ABO + OBC + DBC = 180°.
∴ OBC + half DBC = 90°,
∴ OBC + EBC = 90°,

Hence EBO is a right angle, and EB is tangent to the circle at B.

The triangles BOC and BDC are similar,
The triangles BEC and BCF are similar.

ELEVENTH.

The angle EBC is termed the *deflection* angle, it is equal to half the angle COB, and to the angle CFB, = CBG = GBH = HBI, because they stand upon equal arcs, BC, CG, GH, HI.

PROPOSITION IV.

To lay out a Railroad Circular Curve by Angles of Deflection.

Let the angle of intersection QTC of the two tangents = 112° 40′ = the sum of the two angles TAD, TDA, formed by any connecting line AD; the radius = 200 chains of 100 feet each.

Fig. 8.

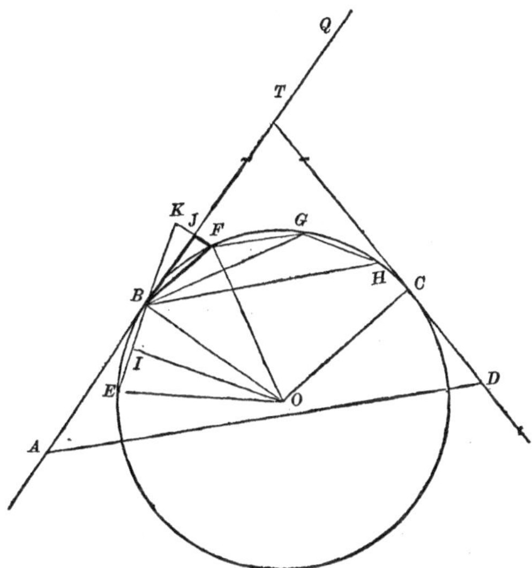

$$BT = TC = OB \times \tan. \tfrac{1}{2}\, QTC.$$

OR, BY LOGARITHMS.

$$\text{Log. } BT = \log. OB + \log. \tan. \tfrac{1}{2}\, QTC.$$

CALCULATION.

	Logarithms.
OB = 200	2·3010300
½ QTC = 56° 20′	10·176476
BT = 300·26..................	2·477506

The tangent BT = 300 chains 26 feet, or 30026 feet. In the

index of the logarithm of BT, 10 was rejected. By transposing the above expression for the tangent BT, we have

$$\text{Log. OB} = \text{log. BT} - \text{log. tan. } \tfrac{1}{2}\text{ QTC.}$$

Hence the radius may be found when the tangent BT and the inclination QTC are known.

The chords EB, BF, FG, &c., are generally a chain in length, whether the chain be one of 100 feet, 50 feet, or 66 feet. In this case I will take the chord BE = BF = FG = GH = &c. = 100 feet.

The angle of deflection JBF = IOB = FBG = GBH, &c., hence,

$$\frac{\text{BI}}{\text{BO}} = \text{Natural sine of IOB} =$$

Angle JBF angle of deflection.

BY LOGARITHMS.

$$\text{Log sin. JBF} = \text{log. BI} - \text{log. BO.}$$

CALCULATION.

	Logarithms.
BI = 50	1·698970
OB = 20000	4·301030
Angle JBF = 0° 8′ 35″ 663	7·397940

Set the theodolite at B, and lay off the deflection angle 8′ 35″·663 from BT, then we have the direction BF, and 100 feet or a chain measured from B to F, will determine F, Measure off the angles FBG, GBH, &c., each equal to 8′ 35″·663, and make BF, FG, GH, &c., each equal 100 feet, and we may determine as many points or stations, F,G, H, &c., in the curve, as we please.

To lay off an angle of 8′ 35″·663 would be a difficult operation, and hence the engineer is obliged to select an angle of deflection that can be set off with ease by the theodolite. When the angle is selected, the radius of the circle EFC, must be increased or diminished, according as the angle 8′ 35″·663 is diminished or increased, and the tangent point B must be taken further from, or nearer to T.

What is the radius BO, when the deflection angle = 9′, (an angle that can be easily measured by a common instrument.)

$$\text{Log. BO} = \text{log. BI} - \text{log. sin. JBF.}$$

CALCULATION.

	Logarithms.
BI = 50	1·698970
JBF = 9′	7·417968
BO = 19098.6 feet	4.281002

When the radius was 20000 feet B was distant 30026 feet from T; now the position of the tangent point B is required when the radius is reduced to 19098·6 feet; the angle of inclination QTC remains unchanged.

Log. BT = log. OB + log. tan. ½ QTC.

CALCULATION.

	Logarithms.
OB = 19098·6 feet	4·281002
½ QTC = 56° 20′....................	10·176476
BT = 28673 feet	4·457478

30026
28673
1353 feet.

So that B, the starting point of the curve, must, in this case, be taken 1353 feet nearer to T, before we commence to lay off the 9′ = JBF; 18′ = JBG; 27″ = JBH, &c.

EXAMPLE.

Let the *deflection* angle JBF = 8′ 20″; the length of the chord BF, = FG = GH, = HL = &c. = 100 feet; and the angle of *inclination* QTC = 112° 40′, as in the last example, it is required to stake out the curve.

Well graduated theodolites divide the minute into three equal parts, each = 20″, hence 8′ 20″ can be measured without difficulty by such instruments.

Log. BT = log. OB + log. tan. ½ QTC.

Log. OB = log. BI — log. sin. JBF.

	Logarithms.
BI = 50 feet	1·698970
JBF = 8′ 20″	7·384544
OB = 20626·5 feet.........................	4·314426

	Logarithms.
OB = 20626·5	4·414426
½ QTC = 56° 20′	10·176476
BT = 30967· feet........................	4·490902

Let B be the point 30967 feet from T, Fig. 9, where the curve commences, make the angles JBF, FBG, GBH, &c., each equal 8′ 20″, and the chords BF, FG, GH, &c., each equal 100 feet, the stations F, G, H, &c., become known.

It generally happens that the whole curve cannot be staked out from the starting point B; then the instrument must be moved to the last point found, and the direction of the tangent at this point determined, and the deflections taken from this last found tangent as in the first case.

Fig. 9.

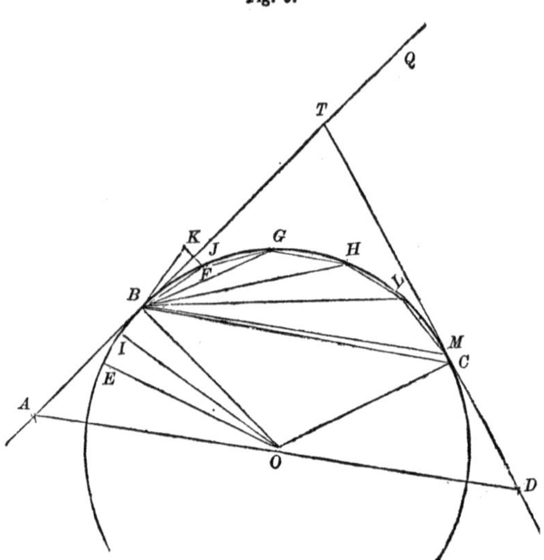

Since each chord EB, BF, FG, GH, &c., subtend angles, EOB, = twice the deflection JBF = in this example to 8′ 20″; and as angle BOC = angle QTC = angle of inclination of the tangents; the number of stations F, G, H, &c., between B and C is readily found by dividing the angle BOC by EOB; or by dividing half BOC by IOB = the tangent deflection.

```
    °  ′                     ′  ″
   56 20                     8 20
   60                        60
  ----                      ----
  3380                      500·
   60                       ----
  ----
5,00)2028,00
     -------
     405·6
```

Hence there will be 405· whole chords and 6-tenths of another over, as MC, Fig. 9. This gives the length of all the chords = 40560 feet, which is sometimes called the length of the curve. This plan of finding the length of the chord of MC is not mathematically true, but in railroad practice the error is too small to be regarded.

To find the length of the arcs BC and BM, and therefore of MC. To radius 1, the length of an arc of one minute = ·00029088820867.

From having the length of an arc of one minute, the following useful table is easily constructed. By this table the length of an arc of any other number of degrees, minutes, &c., is readily found.

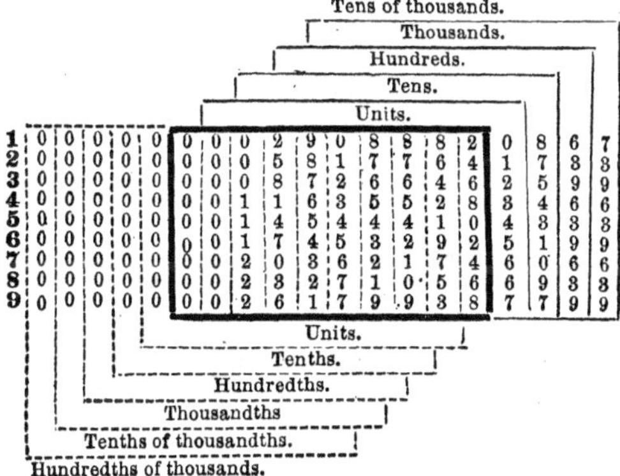

1	0	0	0	0	0	0	0	0	2	9	0	8	8	8	2	0	8	6	7
2	0	0	0	0	0	0	0	0	5	8	1	7	7	6	4	1	7	3	3
3	0	0	0	0	0	0	0	0	8	7	2	6	6	4	6	2	5	9	9
4	0	0	0	0	0	0	0	1	1	6	3	5	5	2	8	3	4	6	6
5	0	0	0	0	0	0	0	1	4	5	4	4	4	1	0	4	3	3	8
6	0	0	0	0	0	0	0	1	7	4	5	3	2	9	2	5	1	9	9
7	0	0	0	0	0	0	0	2	0	3	6	2	1	7	4	6	0	6	6
8	0	0	0	0	0	0	0	2	3	2	7	1	0	5	6	6	9	3	8
9	0	0	0	0	0	0	0	2	6	1	7	9	9	3	8	7	7	9	9

112° 40′ = 6760 minutes.

BY THE TABLE.

6000	1·74532925
700	·20362175
68	·01745329
Arc 112° 40′ to radius 1 =	1·96640429

OB = 20626·5

1·96640429 × 20626·5 = 40560·038087685.

CALCULATION BY LOGARITHMS.

20626·5	4·314426
1·966404	0·293672
40560	4.608098

$\frac{6}{10}$ of 8′ 20″ = 5′ 0″

Length of an arc of 5′ to radius 1 =

```
        ·00145444104
             20626·5
      ---------------
           727220520
          872664624
         290888208
        872664624
      2908882080
      ---------------
      30·000028111560  length to radius 20626.5.
```

These calculations are made in order to show how near such arcs and their chords approximate.

EXAMPLE.

Let the deflection angle JBF = 13°; the distance between station and station as before = 100 feet; and the angle of inclination = 127° = QTC; determine *three* stations from the first or staring point, the tangent to the curve at the last station found, from which lay out the remaining stations. The distance between two points in the straight parts of the railroad to be connected KI = 1230 feet, making angles TIK = 71°, and TKI = 56°, with the tangents TB, TC.

Let B, Fig. 10, be the first tangent point, lay off the angles JBF = FBG = GBH = 13 degrees. Make BF = FG = GH = 100 feet; F, G, H, are station points.

To lay out a tangent PR at the point H, make SHR = the angle PHB = PBH the sum of the deflections up to the station H; in this example the angle SHR = 39° = 3 × 13.

To find the point D, where the curve meets the tangent TC.

$$180^\circ - QTG^\circ = 53^\circ = PTS.$$
$$53^\circ + 39^\circ = 92^\circ = HSR = TBS + BTS.$$
$$HRD = HSR + SHR = 92^\circ + 39^\circ = 131^\circ.$$

$\frac{180^\circ - 131^\circ}{2} = 24^\circ\ 30'$ = the angle RHD = HDR, and hence the point D, where the curve meets the other tangent, becomes known.

$$\text{Or, } 127 - 2 \times 39 = 49$$
$$\frac{49}{2} = 24^\circ\ 30'$$

From Trigonometry we have

$$\text{Sin. } 53^\circ : IK :: \sin. 56^\circ : TI.$$

Fig. 10.

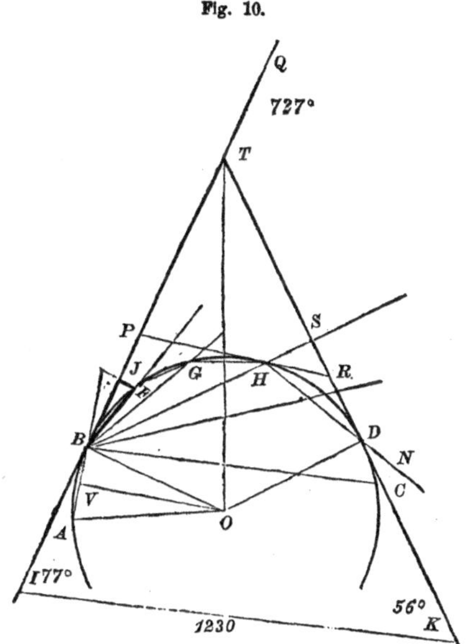

BY LOGARITHMS.

IK, 1230	3·089905
Sin. 56°	9.918574
	13·008479
Sin. 53°	9·902349
TI = 1276·8	3·106130

BA = BF = FG = 100 feet.
Angle BOV = FBG = GBH = 13°.
Log. OB = Log. BV — log. sin. JBF.

BV = 50 feet	1·698970
JBF = 13°	9·352088
OB = 443·5 feet	2·646882

Log. BT = log. OB + log. tan. ½ QTC.

CALCULATION.

	Logarithms.
)B = 443·5 feet	2·646882
½ QTC = 63° 30′	10·302264
BT = 889·5 feet	2·949146

The Engineer having arrived at the station I, and measured the tie line IK = 1230 feet, and the angles TIK, TKI, he calculates

IT = 1276·8 feet.
BT = 889·5 feet.
387·3 = IB.

Before the proposed circular curve is staked out, 387·3 feet are measured from I to B. The radius of the circle = 443·5 feet, this circle is the only one that will connect the two tangents TI, TK, so that a chord of 100 feet will subtend 13° at the circumference or 26° at the centre.

PROPOSITION V.

To lay out a Railroad curve by the Chain only.

Fig. 11.

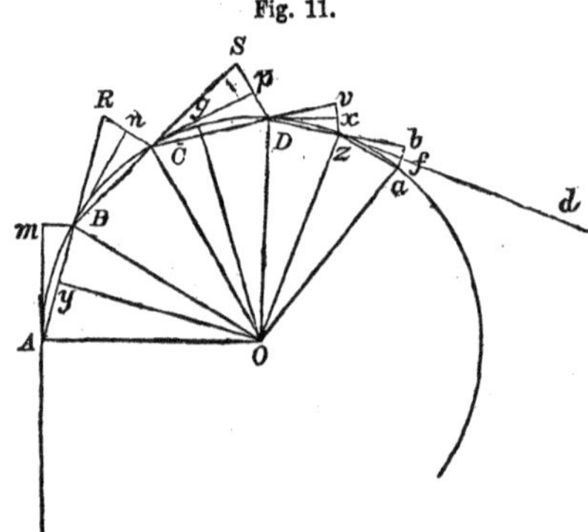

LET the radius AO = 3000 feet, required the tangent deflection *m*B, *n*C, *p*D, &c. and the chord deflection RC, SD, &c. and the length of the chord = 50 feet = AB, = BC, = CD, = CS, = BR.

The triangles BRC and BOC are similar.

These triangles are isosceles, and hence when it is shown that the angle RBC = BOC, their similarity is established.

RBC + the equal angles CBO and OBA = 180°.
BOC + the equal angles CBO and BCO = 180°.
∴ RBC = angle BOC.
∴ OB : BC :: BC : RC that is

$$3000 :: 50 :: 50 : \frac{50 \times 50}{3000} = \frac{5}{6}$$

of a foot = 10 inches.

$$mB = nC = pD = 5 \text{ inches.}$$

$$Am = \sqrt{AB^2 - mB^2} = \sqrt{50^2 - \left(\frac{5}{12}\right)^2}$$

$$\sqrt{600^2 - 5^2} = 599{\cdot}979 \text{ inches.}$$

In this example Am, does not differ from AB much more than the 1-50 of an inch.

Put OB $= r$
BC $= c$
RC $= 2q$; $q = n$C.

$$\therefore 2q = \frac{c^2}{r}$$

It often happens that a curve is in error from not being able to determine the sides of the triangles, AmB, BnC, &c., in whole numbers to suit the particular case. To remedy this, Table I. has been constructed.

Let AB = 1513 feet, links, quarter links, inches or any other convenient measure. In the table opposite 1513, will be found 1512 and 55. Now this table is so formed that 1513 squared is equal to the sum of the squares of 1512 and 55 exactly, and 1513 differs but a unit from 1512; so that mB = 55 feet, &c.. and Am = 1512.

$$1513^2 = 1512^2 + 55^2,$$
$$2289169 = 2286144 + 3025.$$

Any three numbers taken from the table as 1513, 1512, and 55 were taken will form a perfect right angled triangle whose angle mAB, may be as acute as we please. Table I. may be continued to any extent, by the continual addition of 2 to the third line, by continually adding 204, 208, 212, &c., to the second line, the first line is a unit more than the second. Because the triangles AmB, OAy are similar, the radius AO, may be readily found for any set of three numbers taken from the table, thus:

Suppose AB = 365, then
Am = 364, and
mB = 27.

$$365^2 = 364^2 + 27^2, \text{ or,}$$
$$133225 = 132496 + 729.$$

TABLE I.

Chord = c AB, Fig. 11.	Tangent. Am, Fig. 11.	Tangent Deflection, = q Bm, Fig. 11.
5	4	3
13	12	5
25	24	7
41	40	9
61	60	11
85	84	13
113	112	15
145	144	17
181	180	19
221	220	21
265	264	23
313	312	25
365	364	27
421	420	29
481	480	31
545	544	33
613	612	35
685	684	37
761	760	39
841	840	41
925	924	43
1013	1012	45
1105	1104	47
1201	1200	49
1301	1300	51
1405	1404	53
1513	1512	55
1625	1624	57
1741	1740	59
1861	1860	61
1985	1984	63
2113	2112	65
2245	2244	67
2381	2380	69
2521	2520	71
2665	2664	73
2813	2812	75
2965	2964	77
3121	3120	79
3281	3280	81
3445	3444	83
3613	3612	85
3785	3784	87
3961	3960	89
4141	4140	91
4325	4324	93
4513	4512	95
5705	5704	97
5901	5900	99
6101	6100	101

mB : BA : : yA : OA.

$$27 : 365 :: \frac{365}{2} : \frac{365^2}{54} = \text{OA} = 2467\tfrac{7}{54}.\text{ exactly.}$$

The deflection angle mAB is easily found by dividing AB into Bm, or the number in the first column into that in the third column of Table I; the quotient gives the natural sine of the deflection angle mAB.

Let AB = 110·5, and mB = 4·7, be given to find the deflection angle mAB.

$$\frac{4{\cdot}7}{110{\cdot}5} = {\cdot}042534 = \text{natural sine of } 2^\circ\ 26'\frac{77}{291}.$$

What is the deflection angle mAB, when AB = 25, mB = 7, taken from the table.

$$25^2 = 24^2 + 7^2$$
$$625 = 576 + 49.$$

$$\frac{7}{25} = {\cdot}280000 = \text{natural sine of } 16^\circ\ 15'\frac{19}{31}.$$

Required the deflection angle mAB, when the three numbers taken from the table are 6101, 6100, and 101, representing the three sides of the right angled triangles ABm, BCn, CDp, &c.

$$\frac{101}{6101} = {\cdot}016555 = \text{natural sine of } 56'\frac{266}{291}. = \text{angle } m\text{AB}.$$

$$6101^2 = 6100^2 + 101^2.$$

In general there is a short chord, as Dz, at the end of every curve; to such curves the tangent DK is determined, in the same way, as before laid down, between station and station. Make gD = Dz = Dv : divide vz into two equal parts in the point x ; then xD is a tangent to the curve at D. If the straight line of road be continued from z, its direction is found by producing the chord Dz = zb = za. make af = zx, zd is the straight line of road, touching the curve at z.

Put Dz = Dy = za = $c_{,}$; Fig 11.

$2v = 2q_{,}$.

$$. 2q_{,} = \frac{c_{,}^2}{r}, \text{ but it has been before shown that } 2q = \frac{c^2}{r}$$

$$\therefore 2q_{,} : 2q :: \frac{c_{,}^2}{r} : \frac{c^2}{r}.$$

$$q_{,} : q :: c_{,}^2 : c^2$$

$$\therefore q_{,} = q\left(\frac{c_{,}}{c}\right)^2 = zx = af$$

$$\text{Log.}q_{,} = \log. q + 2(\log. c_{,} - \log. c.)$$

EXAMPLE.

Let the two tangents intersect at an angle of 10°, the radius to be any convenient length between 1300 and 1200 feet; find the lengths of the tangents, the length of all the chords that subtend the arc, and a radius, so, that the tangent deflection, the chord, and the tangent to that chord, may form a rational right angled triangle. The deflection angle is to be small, not to exceed 3°.

In one or two trials I find in Table I., that the numbers 1013, 1012, and 45, will answer the required purpose if we take 101·3, 101·2, and 4·5 as feet.

$$4{\cdot}5 : 101{\cdot}3 :: \frac{101{\cdot}3}{2} : \frac{(101{\cdot}3)^2}{9} = 1140\frac{169}{900}$$

the radius exactly; 1140·1877 = the radius nearly.

$$\frac{4{\cdot}5}{101{\cdot}3} = {\cdot}044422 = \text{natural sine of } 2^\circ\ 32'\frac{22}{29}, \text{ the deflection angle.}$$

Tangent = Radius × tan. ½ 10°. (See Fig. 8.)

Log. BT = log. OB + log. tan. ½ QTC.

	Logarithms.
OB = 1140·7877	3·056977
½QTC = 5·0	8·941952
Tangent = 99·75 feet.	1·998929

$$\text{The number of chords} = \frac{5}{2^\circ\ 32'\ \frac{22}{29}} = \frac{300 \times 29}{152 \times 29 + 22} =$$

$$\frac{8700}{4430} = 1{\cdot}964.$$

Fig. 12.

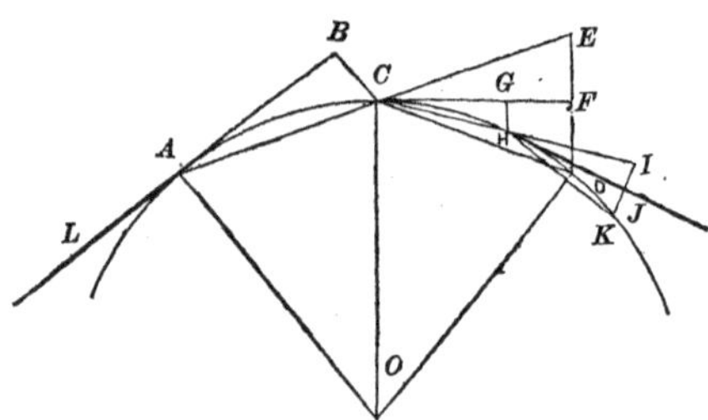

Lay off AB, Fig. 12, equal 101·3 links, BC = 4·5 links, and AB = 101·2 links; the links being a foot in length. ABC is a rational right angled triangle, for

$$(101{\cdot}3)^2 = (101{\cdot}2)^2 + (4{\cdot}5)^2 \text{ exactly.}$$

Let AC be continued to E, make CE = AC = CD = 101.3 feet, and ED = 9 feet, the deflection from the chord AC produced.

Then A, C, D, are points in the circle which touch the tangents or straight lines of road HJ, AL.

But two stations are not required, only 1 station and ·964 of another.

The length of the chord of the whole station between A and C = 101·3 feet, and the chord CH, may be taken = 101·3 × ·964 without sensible error; supposing the arc CH, to be ·964 of the arc AC or CD.

$$CH = 101{\cdot}3 \times {\cdot}964 = 97{\cdot}6532 \text{ feet.}$$

$$c = AC = 101{\cdot}3 \text{ feet.}$$
$$c_{,} = CH = 97{\cdot}6532 \text{ feet.}$$
$$q = BC = 4{\cdot}5 \text{ feet.}$$

$$q_{,} = GH = q\left(\frac{c_{,}}{c}\right)^2, \text{ or by logarithms,}$$

$$\text{Log. } q_{,} = q + 2 \text{ log. } c_{,} - 2 \text{ log. } c.$$

	Logarithms.
q = 4·5	0·653213
2 log. c, ...	3·979374
	4·632587
2 log. c	4·011218
GH = 4·1818	·621369

$$GH = 4\tfrac{2}{11}$$

$$CG = \sqrt{CH^2 - HG^2} = 97{\cdot}56 \text{ feet.}$$

By laying out the curve to D, the next station beyond H, the tangent CF becomes known; measure on CF a distance CG = 97·56 feet, and make GH = 4 2-11 feet: the point H is where curve ends, and where the straight road HJ begins. If the direction of the straight road HJ, be required, continue CH, and make HI = HC = 97·6532 feet = HK;

$$\text{make IK} = 8\tfrac{4}{11} \text{ feet.}$$

K is a point in the curve as well as D, and HJ is a tangent at H, when KJ = JI; and hence HJ is the straight road to connect the curve at H. In this way we can prove whether our work be correct or not; for if the line HJ do not coincide with the given straight line of road the work is not correct.

PROPOSITION VI.

To lay out a Railway Curve by Baker's first method.

Fig. 13.

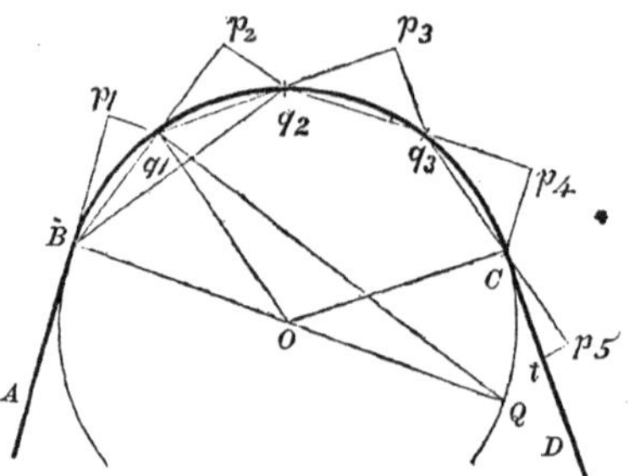

Suppose AB, CD to be two straight or tangental portions of a railroad, it is required to join the points B and C, by a circular curve BC, to which AB, CD, shall be tangents at the points B and C. Let the radius OB be determined by any of the methods that I have laid down, to suit the particular case, and put it $= r$. Complete the semicircle BCQ, prolong BO to Q; join q_1 Q and B q_2. The diagrams upon which the reasoning is carried out, are distorted to make the subject matter clear; for example, Fig. 14 is the true figure, Fig. 12 the corresponding and distorted one.

Fig. 14.

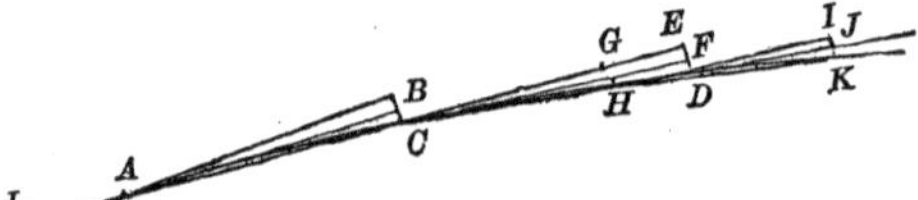

The length of the sine of 3° = ·05234.
The length of the arc of 3° = ·05236.
The length of the tangent of 3° = ·05241.

These lines are nearly of equal length and for less arcs they approach more nearly. Hence, Fig. 13, the tangent BF, the arc BD, and the chord CD, are all very nearly the same length. When as in Fig. 13, the radius is large, and the degrees of the arc small, the length of any chord of a double arc HD does not differ one-fortieth part of a foot from the sum of the chords AC, CD. I shall now return to Fig. 12, to establish the method proposed.

Because Bp_1, Bp_2, Bq_2 are always, in practice, so very small compared with BQ, that they nearly coincide with the curve, and consequently Bp_1 is approximately $= Bq_1 = q_1q_2 =$ half Bq_2.

It has been before shown that angle $p_1Bq_1 =$ angle q_1Bq_2, and hence from the nature of the circle the triangles Bq_1Q, Bp_1q_1, and Bp_2q_2 are similar. Put $Bq_1 = q_1q_2 = c$.

$$\therefore BQ : Bq_1 :: Bq_1 : p_1q_1, \text{ that is,}$$

$$2r : c :: c : p_1q_1 = \frac{c^2}{2r}.$$

$$\text{And } BQ : Bq_1 :: Bq_2 : p_2q_2, \text{ that is,}$$

$$2r : c :: 2c : p_2q_2 = \frac{c_2}{r}.$$

EXAMPLE.

Let $BO = r = 80$ chains, of 66 feet each; then $\frac{1}{2r} = \frac{1}{160}$ of a chain, which being multiplied by 792, the number of inches in each of the chains employed, gives $\frac{792}{160} = 4{\cdot}95$ inches, $=$ the first offset p_1p_1; and

$$4{\cdot}95 \times 2 = 9{\cdot}9 = p_2\ q_2 = p_3\ q_3, \text{ \&c.}$$

Lastly, measure $q_5\ c\ p_5 =$ two chains, the last offset $p_5\ t$ is equal the first offset $p_1\ q_1$.

EXAMPLE.

Let $BO = 75$ chains of 100 feet each: then $\frac{c}{2r} = \frac{100}{150}$ of a foot. $\frac{100}{150} \times 12 = 8$ inches $= p_1q_1$ or the last offset $p_5\ t$.

$$p_2q_2 = 16 \text{ inches} = p_3\ q_3, \text{ \&c.}$$

PROPOSITION VII.

To lay out a railroad curve by ordinates or offsets from its tangents.

CASE I.

When the length of the curve is less than one quarter of its radius.

Let BC be the curve, AB and CD tangents at B and C ranged to meet at T. The radius r = OB, may be determined by any method

Fig. 15.

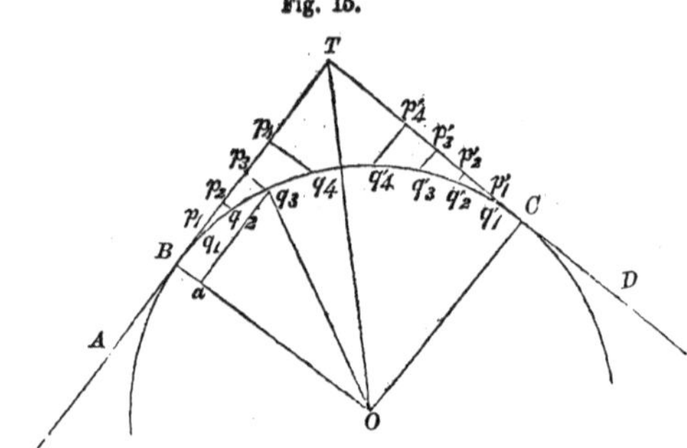

previously laid down. Measure on BT the distance $Bp_1 =$ one chain say of 100 links, each a foot in length. A 50 foot chain of 50 links may be employed; or Gunter's chain of 100 links and 66 feet long may be preferred by some engineers. $p_1p_2 = 1$ chain; $p_2p_3 = 1$ chain, &c.

$$\text{Make } p_1q^1 = \frac{1}{2r} \text{ of a chain.}$$

$$p_2q_2 = \frac{2^2}{2r} \text{ of a chain.}$$

$$p_3q_3 = \frac{3^2}{2r} \text{ of a chain.}$$

$$p_4q_4 = \frac{4^2}{2r} \text{ of a chain.}$$

&c., &c.,

or, what amounts to the same thing, the 2d, 3rd, 4th, &c., offsets, are respectively 2^2, 3^2, 4^2, &c., times the first offset, or 4, 9, 16, &c., times the first.

Having laid out the offsets in this manner, till the last one, which suppose to be p_4q_4, is either within, or very little more than, a chain from T, make $p_4'T, = p_4T$, and lay out the same offsets in an inverted order on TC, as were laid down on BT, that is, beginning with the greatest first and ending with the least.

EXAMPLE.

Let the radius of the curve be 60 chains, each of 50 links, a foot each.

Chain employed = 600 inches long.

$$p_1q_1 = \frac{1}{2 \times 60} \times 600 = 5 \text{ inches}$$

$p_2q_2 = 5 \times 4 = 20$ inches.
$p_3q_3 = 5 \times 9 = 45$ inches.
$p_4q_4 = 5 \times 16 = 80$ inches.
&c. = &c.

Example with a 66 foot chain.

Let the radius of the curve be 80 of these chains, then $p_1q_1 = \frac{1}{2 \times 80} \times 792 = 4{\cdot}95$ inches; $p_2q_2 = 4{\cdot}95 \times 4 =$ inches 19·8 $p^3q_3 = 4{\cdot}95 \times 9 = 44{\cdot}55$ inches, &c.

The truth of the rule just applied may be established as follows :—Draw the radius Oq_3, and q_3 a perpendicular to OB, which put $= r$.

$$\text{Put } n = aq_3 = Bp_3.$$

$$\text{Then } Oa^2 = \sqrt{r^2 - n^2}, \text{ and}$$

$$Ba = p_3q_4 = OB - Oa = r - \sqrt{r^2 - n^2}.$$

Now if n_2 be taken successively $= 1^2, 2^2, 3^2$, &c., chains, the values of

$$p_1q_1 ;\ p_2q_2 ;\ p_3q_3 ;\ \&c.$$

will be respectively

$$r - \sqrt{r^2 - 1^2} ;$$

$$r - \sqrt{r^2 - 2^2} ;$$

$$r - \sqrt{r^2 - 3^2} ;$$

$$r - \sqrt{r^2 - 4^2} ;$$

&c.

But since, $1^2, 2^2, 3^2$, &c. are very small compared with r_2, the values of $\sqrt{r^2 - 1^2}$, $\sqrt{r^2 - 2^2}$, &c. may be taken respectively

$r - \frac{1^2}{2r}$; $r - \frac{2^2}{2r}$, &c. without material error. ∴

$$p_1q_1 = r - \left(r - \frac{1^2}{2r}\right) = \frac{1^2}{2r},$$

$$p_2q_2 = r - \left(r - \frac{2^2}{2r}\right) = \frac{2^2}{2r},$$

$$p_3q_3 = r - \left(r - \frac{3^2}{2r}\right) = \frac{3^2}{2r}, \&c.$$

When the length of the curve exceeds $\frac{1}{4}$ of the radius, only 5 or 6 of the approximate offsets ought to be taken in this manner. and the remainder, say, commencing with $p_7 q_7$ ought to be taken $= r - \sqrt{r^2 - 7^2}$, and so on.

Such expressions as $\sqrt{r^2 - 7^2}$ are easily and correctly worked by logarithms, for Logs.

$$\sqrt{r^2 - 7^2} = \frac{1}{2}\left\{\log.(r+7) + \log.(r-7)\right\} \text{ or } \frac{r - \sqrt{r^2 - 7^2}}{r}$$

$= 1 - \sqrt{1 - \overline{7}^2}$, hence, if an arc or angle be found whose sine $= \frac{7^2}{r}$, the versine of this arc taken from a table and multiplied by r, we have $p_7 q_7$, and so of other ordinales.

When the Curve is any given length.

Since it is inconvenient in practice to have the offsets much greater than 2 chains in length ; if, therefore, the curve be a long one, the offsets may be confined within proper limits by dividing the curve into two or more parts, and introducing one or more additional tangents. In Fig. 16, the curve BC is divided into two

Fig. 16.

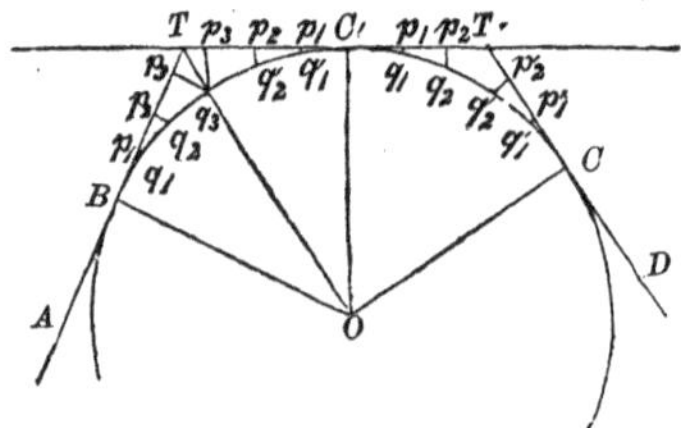

parts at C′, at which point the tangent TC′T′ is drawn, meeting the tangents BT, T′C in T and T′, the tangents BT, TC′ being each taken $= Bp_3 + \frac{1}{8} p_3 q_3$, in which Bp_3 is taken the nearest whole number of chains to $\frac{1}{8}$ BO, and $p_3 q_3$ is the offset corresponding to Bp_3. It is easily seen that we propose to lay off an arc whose tangent is not to exceed by any great length the $\frac{1}{8}$ part of the radius, so that the offsets may be found with ease and accuracy ; and it may be necessary to show how

$$BT \text{ or } TC' = Bp_3 + \tfrac{1}{8} p_3 q_3.$$

This is easily done, for the triangles OBT, and $q_3 p_3 T$ are similar,

$$\therefore OB : Bp_3 + p_3T :: p_3q_3 : p_3T ;$$

But since it has been agreed upon $Bp_3 = \frac{1}{8}$ OB or the nearest whole number to it, and since p_3T must always be very small compared with Bp_3, we have,

$$OB : \tfrac{1}{8}OB \ (= BT \text{ nearly}) :: p_3q_3 : p_3T.$$
$$\therefore \text{ in such cases } p_3T = \tfrac{1}{8}\, p_3q_3.$$

By thus taking the length of BT, TC′, the last offset on BT and the first one on TC′ will meet at q_3, the middle point of BC′ and hence the distances of the ends of the offsets, that form the curve, will be sufficiently near equality.

Again, BT : BO :: Radius : tan. BTO = ½ BTC′ ∴ the direction of the new tangent TC′T′ becomes known.

Having made $Tp_3 = Tp_3'$; $p_2q_2 = p_2'q_2'$, &c., in an inverted order on TC′; invert the order of the offsets a second time on C′T′; and a third time on T′C.

EXAMPLE.

Let BO = 240 chains of 66 feet each, then $Bp_3 = \frac{1}{8}BO$, conventionally, or agreed that it should be so. $p_3q_3 = \dfrac{(Bp_3)^2}{2 \times BO} = \dfrac{30^2}{240} = 1{\cdot}875$ chains.

$$\therefore BT = TC' = Bp_3 + \tfrac{1}{8}\, p_3q_3 = 30 + \frac{1{\cdot}875}{8} =$$

30·2344 chains.

TRIGONOMETRICAL CALCULATION.

	Logarithms.
As 30·2344 = BT	1·480502
: 240 = OB	2·380211
:: Radius	10·000000
: tan. ½BTC′ = 82° 49′$\frac{1}{5}$	10·899709

82° 49′$\frac{1}{5}$
2

BTC′ = 165 38$\frac{2}{5}$

It is unnecessary to find the angle C′T′C when the position of the tangent TT′ is known.

To lay off an angle of degrees, &c. on the ground with the chain only.

This may be done very correctly by employing continued fractions.

EXAMPLE.

Let it be required to lay off the angle 165° 38$\frac{2}{5}$, with the chain only.

$$\begin{array}{rr} 180^\circ & 0' \\ 165 & 38\frac{2}{5} \\ \hline 14 & 21\frac{3}{5} \end{array}$$

Natural sine of $14^\circ\ 21'\frac{3}{5} = \cdot248013$

Natural cosine of $14^\circ\ 21\frac{3}{5} = \cdot968757$

248013)968757(3 number used.
744039

224719)248013(1 number used.
224719

23294)224719(9 number used.
209646

15073)23294(1
15073

8221

8221)15073(1
8221

6852)8221(1
6852

1369)6852(5
6845

7, &c., &c.

$$\cfrac{1}{3+\cfrac{1}{1+\cfrac{1}{9}}} = \frac{10}{39} \text{ nearly.}$$

$$\cfrac{1}{3+\cfrac{1}{1+\cfrac{1}{9+\cfrac{1}{1+\cfrac{1}{1}}}}} = \frac{21}{82} \text{ more nearly.}$$

$$\cfrac{1}{3+\cfrac{1}{1+\cfrac{1}{9+\cfrac{1}{1+\cfrac{1}{1+\cfrac{1}{1+\cfrac{1}{5}}}}}}} = \frac{181}{707} \text{ more nearly.}$$

·248013 : .968757 : : 10 : 39 nearly.
248013 : 968757 : : 21 ·: 82 more nearly.
248013 : 968757 : : 181 : 707 more nearly.

Yet, and so on, by summing the continued fractions we may find whole numbers that will approach as near as we please to the ratio existing between the sine and cosine of the angle to be laid out.

$$181 \times \cdot 968757 = 175 \cdot 345017.$$
$$707 \times \cdot 248013 = 175 \cdot 345191.$$

This shows the accuracy of the numbers worked out. I have made great use of this rule in the practice of Engineering, both civil and mechanical, and lay it before the public in this work for the first time.

Fig. 17.

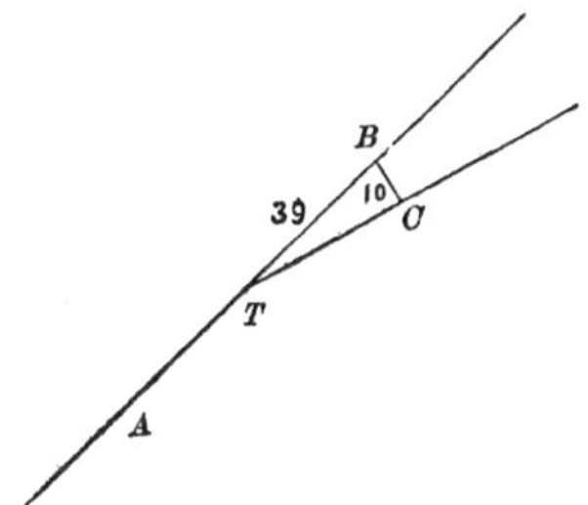

Fig. 17, make TB = 39 links, and BC perpendicular to it = 10 links, the angle BTC = 14° 21′$\frac{3}{5}$ nearly.

Fig. 18.

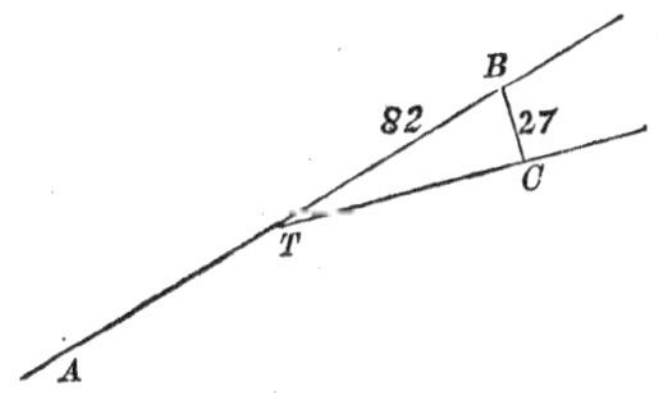

Fig. 18, make TB = 82 links, and BC perpendicular to it = 21 links, then the angle BTC = 14° 21′$\frac{3}{5}$ more nearly.

Fig. 19, make TB = 707 links, and BC perpendicular to it = 181 links, then the angle BTC may be said to be exactly = 14° 21′$\frac{3}{5}$ and ATC = 165° 38′$\frac{2}{5}$. This may be tested by dividing 181

Fig. 19.

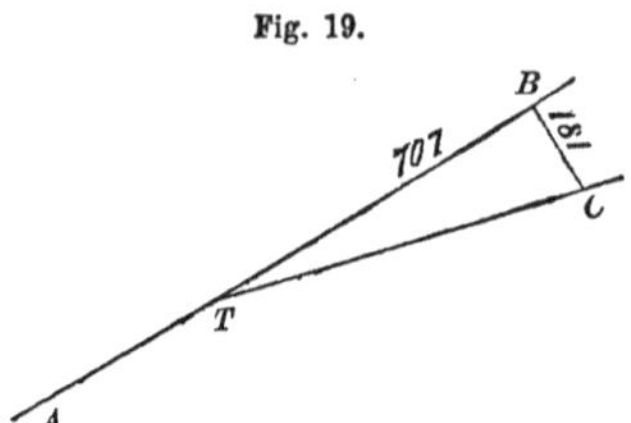

by 707, and if the quotient gives the natural tangent of $14^\circ\ 21'\frac{3}{5}$ the work is correct.

$$\frac{181}{707} = \cdot 256011 = \text{natural tangent of } 14^\circ\ 21'\tfrac{3}{5}.$$

By this method angles may be laid off by a chain, tape or other measures divided into equal parts.

The first offset p_1q_1, Fig. 16, $= 1{\cdot}65$ inches, or,

$$\frac{1}{2 \times 240} \times 792 = \frac{792}{480} = 1{\cdot}65 \text{ in.}$$

from which the other 29 offsets may be found by multiplying by 4, 9, 16, 25, &c. p_3q_3 is in the position of the 30th offset, the length of which $= 1{\cdot}65 \times 900 = 1485{\cdot}00$ inches $= 123$ feet 9 inches.

If the curve be one of about 80 chains radius, and between 32 and 38 chains in length, then to avoid the trouble of adding another tangent, the offsets beyond the eighth must be calculated from the following formulas:

$$p_9q_9 = r - \sqrt{r^2 - 9^2};$$
$$p_{10}q_{10} = r - \sqrt{r^2 - 10^2};\ \&c., \&c.$$
$$\therefore p_{30}q_{30} = r - \sqrt{r^2 - 30^2}.$$

In the example just given $r = 240$ chains.

$$240^2 = 57600$$
$$30^2 = 900$$
$$\sqrt{56700} = 238{\cdot}117618$$

From 240
Take 238·117618

$p_{30}q_{30} =$ 1·882382 chains.
792

3764764
16941438
13176674

$p_{30}q_{30} =$ 1490·846544 inches.

By approximation we made $p_{30}q_{30} = 1485$ inches, which differs from 1491 but 6 inches. So that the accuracy of the approximate rule is very great.

If the radius of the curve be 160 chains, and its length about 50 or 60 chains, the offsets must be calculated as above, beyond the 15th.

$$\text{The formula } \frac{1^2}{2r},\ \frac{2^2}{2r},\ \frac{3^2}{2r},\ \&c.$$

being a very near approximation to the true lengths of the offsets from the tangent, within the limits assigned, but beyond these limits the errors of the offsets begin gradually to augment, till they become too considerable to be overlooked.

EXAMPLE.

Suppose a chain of 100 feet and 100 links is used, the radius of the curve = 150 chains, or 15,000 feet, what are the lengths of the 1st, 2nd, 3rd, and 10th offsets in inches; the 10th to be found both approximately and correctly?

$$\frac{1^2}{300} \times 1200 = 4 \text{ inches} = p_1 q_1$$

$$\frac{2^2}{300} \times 1200 = 16 \text{ inches} = p_2 q_3$$

$$\frac{3^2}{300} \times 1200 = 36 \text{ inches} = p_3 q_3$$

$$\frac{10^2}{300} \times 1200 = 400 \text{ inches} = p_{10} q_{10}$$

$$p_{10} q_{10} = r - \sqrt{r^2 - 10^2} \text{ chains.}$$

$$150^2 = 22500$$
$$10^2 = 100$$
$$\sqrt{22400} = 149{\cdot}666295$$

From 150·
Take 149·666295

$$p_{10} q_{10} = {\cdot}333705 \text{ chains}$$
1200

$$p_{10} q_{10} = 400{\cdot}446000 \text{ inches.}$$

Hence the tenth offset does not differ from the truth ½ an inch when found by approximation.

EXAMPLE.

Suppose a 50 foot chain of 50 links is employed, the radius of the curve 700 such chains, what are the lengths of the 1st, 2nd,

3rd offsets approximately, and the 20th approximately and correctly?

$$\frac{1^2}{1400} \times 600 = \tfrac{3}{7} \text{ inch} = p_1 q_1$$

$$\frac{2^2}{1400} \times 600 = 1\tfrac{5}{7} \text{ inch} = p_2 q_2$$

$$\frac{3^2}{1400} \times 600 = 3\tfrac{6}{7} \text{ inches} = p_3 q_3$$

$$\frac{20^2}{1400} \times 600 = 171\tfrac{3}{7} \text{ inches} = p_{20} q_{20}$$

$$p_{20}\, q_{20} = r - \sqrt{r^2 - 20^2} \text{ chains.}$$

$$700^2 = 490000$$
$$20^2 = 400$$
$$489600 = 699{\cdot}714228$$

From 700·
Take 699·714228

$p_{20}\, q_{20} =$ ·285772 chains.
600

171·463200 inches.
Found by approx. 171·428571

·034629

Hence the approximate value of $p_{20}\, q_{20}$ is extremely near the truth.

PROPOSITION VIII.

To lay out a Railroad curve by two Theodolites when obstructions prevent the use of the Chain.

In the third book of Euclid it is shown that the angles ACB and ADB (Fig. 20) are together = 180°. All angles in the same segment are equal to one another: angle ACB = AEB, and angle ADB = AFB.

The angle at the centre AOB = twice the angle at the circumference ADB or AFB.

$$\therefore \text{Angle ACB} + \text{half AOB} = 180^\circ$$

The angle CAB + ACB + CBA = 180° also

$$\therefore \text{ACB} + \text{CAB} + \text{CBA} = \text{ACB} + \tfrac{1}{2}\text{AOB}$$
$$\therefore \text{CAB} + \text{CBA} = \tfrac{1}{2}\text{AOB}.$$

Fig. 20.

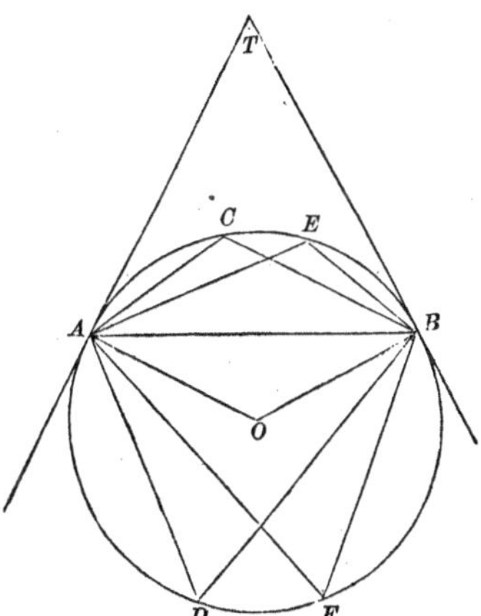

This is the principal property made use of in establishing the following method of staking out the curve.

The position of the tangents PA and RZ, the radius AO, and the angle AOZ, being determined, join (Fig. 21) AZ, and take

Fig. 21.

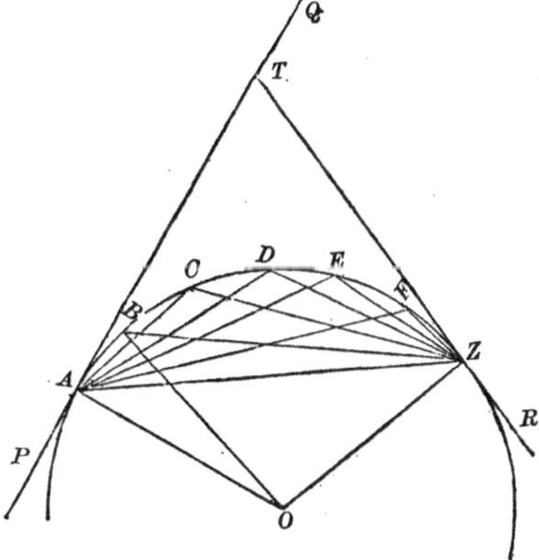

several equi-distant points B, C, D, &c. in the curve, from which points draw lines to A and Z. Put c = AB, BC or chord CD, &c. and r = AO, the radius.

Take BZA = an angle whose sine is $\frac{c}{2r}$, or, because c is in practice usually = 1 chain, whether that chain be 100, 50, or 66 feet in length. Then r must be taken in chains also, and BZA = an angle whose sine $= \frac{1}{2r}$.

The several angles may be arranged and determined thus:—

$$\text{BZA} = \text{Angle to sine } \frac{1}{2r} \text{ and BAZ} = \tfrac{1}{2}\text{AOZ} - \text{BZA}.$$

CZA = 2BZA and CAZ = ½AOZ — 2BZA.
DZA = 3BZA and DAZ = ½AOZ — 3BZA.
EZA = 4BZA and EAZ = ½AOZ — 4BZA.
&c. = &c. &c. = &c.

Consequently if a theodolite be fixed at A and Z, and the angles BZA, and BAZ, be taken at the same time, the intersection of AB and BZ will give the point B. In the same manner by taking the angles CZA, and CAZ, the intersections of AC and ZC will give the point C.

EXAMPLE.

Let the radius AO = 80 chains, and the angle between the tangents PT; RT; = 106° = ATZ. Then angle

$$\text{AOZ} = 180° - 100° = 80°.$$
$$\text{half AOZ} = 40°.$$
$$\frac{1}{2r} = \frac{1}{160} = \cdot006250.$$

·006250 = Natural sine of 0° 21′ 29″ = BZA.
once 0° 21′ 29″ = 0° 21′ 29″. = BZA.
twice 0° 21′ 29″ = 0° 42′ 58″. = CZA.
3 times 0° 21′ 29″ = 1° 4′ 27″. = DZA.
4 times 0° 21′ 29″ = 1° 25′ 56″. = EZA.
&c. = &c.

BAZ = ½AOZ — BZA = 39° 38′ 31″.
CAZ = ½AOZ — CZA = 39° 17′ 2″.
DAZ = ½AOZ — DZA = 38° 55′ 33″.
EAZ = ½AOZ — EZA = 38° 34′ 4″.
&c. = &c. = &c.

This list of angles must be continued till a multiple of BZA = in this example to 21′ 29″, can no longer be taken from half AOZ.

PROPOSITION IX.

To lay out a Railroad Curve by means of Ordinates or Offsets from the Chord or Chords.

Let BEC be a portion, or the whole of a curve of a railroad; AT, TF, CD, tangents to the curve at the points B, E, C; O the centre; OB the radius; BE, EC chords of the curve. This plan is employed chiefly between station and station, when the rails are being put down. Upon this view of the proposition, BE is supposed to represent 100 feet, or one chain of that length. It often happens that a 50-feet chain of 50 links is employed. The Fig. 22, as in other cases, is an extravaganza for the sake of clearness.

Fig. 22.

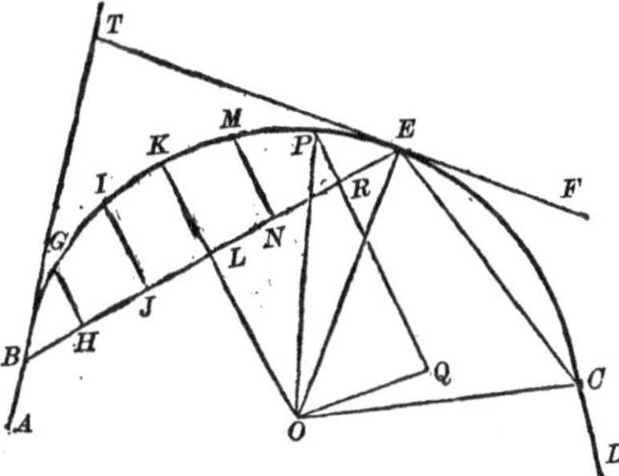

Let the radius OK bisect the curve and chord in K and L. Since the angle BOK = TBE, hence the position of the chord is determined, as the angle BOK is supposed given. When the curve is large, put OB = r, OL = s, and the half chord BL = n, chains.

$$\sqrt{r^2 - n^2} = s$$

$$PR = \sqrt{r^2 - (n-1)^2} - s,$$

$$MN = \sqrt{r^2 - (n-2)^2} - s,$$

$$\text{The 111th ordinate} = \sqrt{r^2 - (n-m)^2} - s.$$

After reaching K, or the middle point of the curve, the same offsets are repeated in an inverted order till the curve shall have been set out to E. The same operation may be repeated as often as necessary, till the whole curve be completed, observing to make the angle BEC = twice the complement of the angle TBE, which has been already found.

If the last chord, which suppose to be EC, be less than the preceding chords or chord, the angle OEC must be found and added to the known angle OEB, which will give the angle BEC, showing the direction of the chord EC.

EXAMPLE.

Let it be required to connect two straight portions of a railroad UP, VW, that make with one another an angle PSV $= 112°$. Say it is convenient to find 8 ordinates to the arc, that its chord be 1000 or thereabouts, so that 8 may be contained in it evenly. (It might be taken 1008, 1016, 1024, &c., and yet the object in view attained.)

The radius AO is to be between 600 and 800 feet in length, but the right angled triangle AKO must be expressed by rational numbers. These restrictions are made so that the ordinates may be found by a table of the squares, square roots, &c. of numbers, by mere inspection.

$$KL = LM = MN = NB = 125 \text{ feet}$$
$$AK = 500 \text{ feet.}$$

Divide 500 by 2 and we have 250, take away two numbers whose product is 250, as 25 and 10; the sum of the squares of these numbers $= 725$, that is

$$25^2 + 10^2 = 725 = AO.$$

This number will suit our purpose to render the triangle AKO rational, for,

$$OK = \sqrt{725^2 - 500^2} = 525.$$

The square of 725 taken from table, ...	=	525625
500^2 taken from the table, ...	=	250000
525 squared from table, ...	=	275625

Hence we find, without calculation, the sides of the right angled triangle AKO, with OA, the radius of the required length, as it was necessary to have it range between 600 and 800 feet.

$$\text{Also, } 250 = 50 \times 5$$
$$250 = 125 \times 2$$
$$250 = 16 \times 15\tfrac{5}{8}$$

$$\left.\begin{array}{l} 50^2 + 5^2 = 2525 \\ 125^2 + 2^2 = 15629 \\ 16^2 + (15\tfrac{5}{8})^2 = 500\tfrac{9}{64} \end{array}\right\} \text{Any of these numbers may be taken for AO, and render the triangle AKO rational.}$$

$$500^2 = 250000 = \frac{1024000000}{64^2} \text{ take}$$

$$(500\tfrac{9}{64})^2 = \frac{1024576081}{64^2} \text{ from.}$$

The remainder $= \dfrac{576081}{(64)^2}$ the square root of which $= \dfrac{759}{64} = OK,$

Fig. 23.

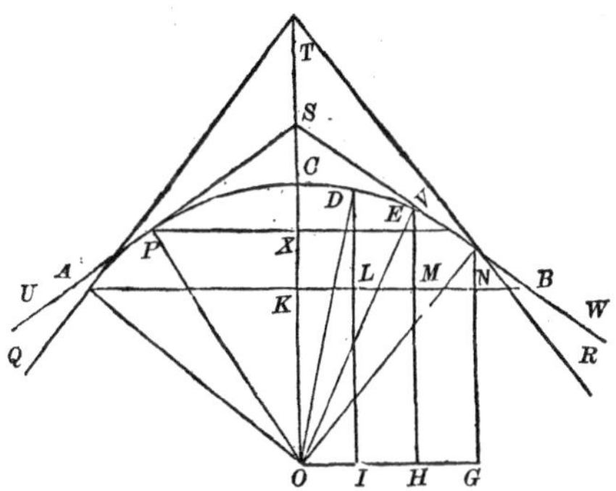

which is a rational number, although not a whole number ; such numbers are often of great use in practice, and curtail much labor. Hence, to sum up this particular case, in finding the sides of the triangle AOK between limits, and so that these sides may be expressed by rational numbers, we may repeat

$$AK = 500 \text{ feet}$$
$$AO = 500\tfrac{9}{64} \text{ "}$$
$$OK = 11\tfrac{55}{64} \text{ "}$$
$$(500\tfrac{9}{64})^2 = (500)^2 + (11\tfrac{55}{64})^2, \text{ exactly.}$$

Having shown the numerous ways in which the triangle AKO may be made rational without changing the length of AK, I will proceed to the particular example before us, in which we have taken

$$AO = 725 \text{ feet,}$$
$$KO = 525 \text{ feet, } AK = 500.$$

$$OG = KN = 500 - 125 = 375 \text{ feet.}$$

$OF^2 = 725^2 = 525625$ taken from table.
$OG^2 = 375^2 = 140625$ taken from table.

385000

Nearest sq. in tab. = 384400 = $(620)^2$

620 × 2 = 1240) 6000 (·483
4960

10400
9920

∴ FG = 620,483 4800
3720

By dividing the remainder 600 by twice 620, the rest of FG is determined with great ease and accuracy.

$$620{\cdot}483 - 525 = 95{\cdot}483 = FN.$$

$$OH = KM = 500 - 250 = 250 \text{ feet.}$$

$$OE^2 = (725)^2 = 525625, \text{ taken tab. of squares.}$$
$$OH^2 = (250)^2 = 62500, \text{ from table of sqs.}$$

```
                 463125
     (680)² =    462400  nearest sq. in tab.
Remainder           725

          680 × 2 = 1360.

       1360)  7250  (·533
              6800
               4500
               4080
                4200
                4080
       EH =     680·533
```

$$680{\cdot}533 - 525 = 155{\cdot}533 = EM.$$

Again,—

$$OI = KL = 500 - 375 = 125 \text{ feet.}$$

$$OD^2 = 725^2 = 525625, \text{ taken from tab. of sqrs.}$$
$$OI^2 = 125^2 = 15625, \text{ taken from table.}$$

```
                   510000
Nearest sq. =      509796  = 714²
       1428)        2040   (714·142 = ID.
                    1428
                     6120
                     5712
                      4080
                      2856
714·142 — 525 =    189·142 = LD.
```

In this way ordinates are calculated with great rapidity. In practice I have employed this method with great success. It was invented by me (Oliver Byrne), the author of the present work. Although it appears rather long, from the length of the explanation, yet offsets may be thus found almost as easily as by inspecting a table. The more extensive the table of the squares of numbers employed the better.

Since the tangents at A and B do not coincide with the straight portions of the road UP, VW, it is necessary to lay down the chord AB in such a place that the curve will touch US and VS. The points P and V are both in the curve and the tangents or straight portions of the road.

OP = 725, angle PSO = 56°.

POS = 90° — 56° = 34° = angle SPX.

Sin. 56° : OP : : sin. 34° : PS.

Hence the points P and V are known.

Sin. 56° : OP : : sin. 90° : SO.

SO — CO = SC, therefore, the position of C is found.

CK = CO — OK,

CK = 725 — 525 = 200.

Hence the position of the chord AKB is determined, so that the arc APCVB laid out by ordinates touch the straight parts of the road at P and V.

EXAMPLE.

Let it be required to connect two straight portions of a railroad UP and VW, that make with one another an angle PTV = 123°. Say that 19 ordinates are required to the arc, that is, AB is to be divided into 20 equal parts. The chord AB is to be about 1000 feet, the radius OP to be between 800 and 900 feet long, so that the right angled triangle AOK, may be expressed by rational numbers. These restrictions are made as in the last example, so that the ordinates may be found by inspecting a table of squares, cubes, of numbers. Commence by laying off the 4th ordinate.

Let AB = 1008 feet, then ½ KB = 252, a number that can be divided by 2, 3, 4, 5, 6, 7, and 9.

$$2 \times 126 = 252$$
$$3 \times 84 = 252$$
$$4 \times 63 = 252$$
$$6 \times 36 = 252$$
$$7 \times 36 = 252$$
$$9 \times 28 = 252$$

We have shown that fractions may be employed with effect, and yet have the triangle AOK rational; for example, $8 \times 31\frac{1}{2}$ are numbers that may be used, when the radius may be $1056\frac{1}{4}$ for

$$8 \times 31\tfrac{1}{2} = 252$$
$$\text{and } 8^2 + (31\tfrac{1}{2})^2 = 1056\tfrac{1}{4}.$$

The numbers 2 and 126 are to be used when the radius = 15880 feet.

But the numbers that will suit the case in question = $9^2 + 28^2 = 865 = AO$.

$$\text{then } OK = 28^2 - 9^2 = 703 \text{ feet,}$$
$$\text{and } KS = 865 - 703 = 162 \text{ feet.}$$
$$AK^2 + KO^2 = AO^2 \text{ in numbers,}$$
$$504^2 + 703^2 = 865^2.$$

This may be soon tested by taking the squares of these numbers from the table of squares, square roots, &c., of numbers:

$$504^2. = 254016$$
$$703^2. = 494209$$
$$865^2 = 748225$$

Angle PTO = 61° 30′;
" TOP = 90° — 61° 30′ = 28° 30′.
" TPX = 28° 30′ also.

	Logarithms.
Sin. 61° 30′ = PTO, ...	9·943899
: 865 = PO.	2·937016
: : sin. 26° 30′ = POT, ...	9·678663
: PT = 469·66, ...	2·671780

TV = PT = 469·66 feet.

Sin. 61° 30′ = PTO, .:	9·943899
: 865 = PO, ...	2·937016
: : sin. 90°, = TPO, ...	10·000000
: TO = 984·28, ...	2·993117

TO — SO = ST, that is,
984·28 — 865 = 119·28.

Hence the distance of S from T is known, and therefore the centre point of the arc is known. The point K may be staked out also for

TK = SK + TS, that is
= 162 + 119·28 = 281·28 feet.

$\frac{1}{10}$ of AB = 50·4 feet = I to B
IV to B = 201·6
IV to K = 302·4 = OG.

From the table of squares, cubes, &c.,

$OV^2 = 865^2 =$ ————	748225
$OG^2 = 301{\cdot}4^2 = 4 \times (150{\cdot}7)^2 =$	90841·96
	657383·04
810^2, nearest tab. sq.	656100·
$2 \times 810 = 1620)$	12830 (·792
	11340
	14904
	14580
	3240
	3240

VG = 810·792
Ordinate IV = 810·792 — 703
= 107·792 feet.

B to V = 252.

$865^2 = 748225$
$252^2 = 63504$

684721

$(827)^2 = 683929$

792

```
1654)  7920  (·478
       6616
       13040
       11578
        14629
        13232
```

827·478 — 703 = 124·478

Ordinate V = 124·478 feet.

If the operator possesses a copy of the "Calculator's Constant Companion," by Oliver Byrne, the author of the present work, the division to find the decimal part of the ordinate as ·478, may be performed without labor, or by mere inspection. In what follows I will not set down any unnecessary figures, or repeat any of the cumbersome explanations employed to make the work clear.

B to VI = 201·6

201·6 cannot be got in the table used, but 100·8 can.

$865^2 = 748225$
$(201{\cdot}6)^2 = 40642{\cdot}56 = (100{\cdot}8)^2 \times 4$

707582·44

$841^2 = 707281$

```
1682)   301·44  (·179
        168 2
        13324
        11774
         15500
         15138
```

841·179
703·

138·179 = ordinate VI.

From B to VII. = 151·2 feet.

$865^2 = 748225·$
$(151·2)^2 = 22952·25$

725272·75

$(851)^2 = 724201·$

1702 1071·75 (·629
10212

5055
3404

16510
15318

851·629
703·

148·629 = ordinate VII.

From B to VIII = 100·8 feet,

$865^2 = 748225·$
$(100·8)^2 = 10160·64$

738064·36

$859^2 = 737881·$

1718) 183·36 (106
1718

11560
10308

859·106
703.

156·106 = ordinate VIII.

From B to IX = 50·4

$865^2 = 748225·$
$50·4^2 = 2540·16$

745684·84

$863^2 = 744769·$

1726 915·84(·530
8630

5284
5178

863·53
703

160·53 = ordinate IX.

By these calculations we have determined 13 ordinates; namely the centre one and six at each side of it, besides the points P and V are in the curve. Hence 15 points are known. By no known method can so many ordinates be found by so few calculations.

EXAMPLE.

It is required to connect two straight portions of a railroad UP and VW, in the triangle UTW, UT = 240·3, UZ = ZW = 240 feet, and TZ = 12 feet; these measurements being taken, angular measurements are not required. AB is to be divided into 56 equal parts by 57 ordinates, the length of AB is to be about 400 feet. The radius OP is required to be about half a mile, but the right angled triangle, AKO, is to be expressed by rational numbers.

Fig. 24.

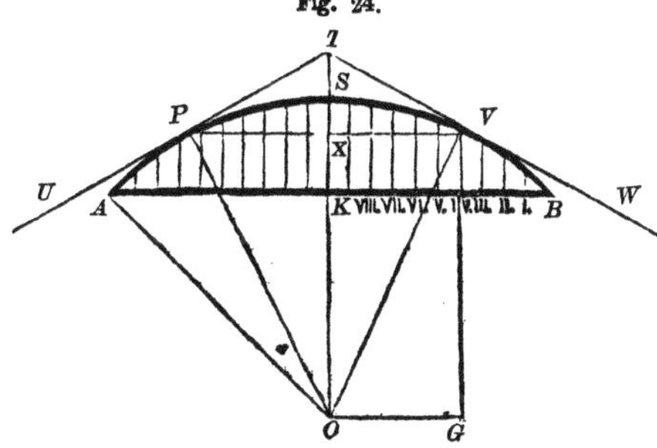

216 = half of AB, may be divided without remainder by 28. The half 216 = 108.

$$54 \times 2 = 108.$$
$$36 \times 3 = 108.$$
$$27 \times 4 = 108.$$
$$18 \times 6 = 108.$$
$$13\tfrac{1}{2} \times 8 = 108.$$
$$12 \times 9 - 108.$$

$54^2 + 2^2$ 2920 feet, which is a little over half a mile = 2640 feet.

$$54^2 + 2^2 = 2920^2 = 8526400$$
$$\text{Twice } (54 \times 2) = 216^2 = 46656$$
$$54^2 - 2^8 = 2912^2 = 8479744.$$

$$AO = 2920 = OP.$$
$$AK = 216.$$
$$OK = 2912.$$
$$\therefore SX = 8 = 2920 - 2912.$$

The three triangles OPT, PXT, UZT are similar.

UZ : ZT : : OP : PT, that is,

240 : 12 : : 2920 : 146 = PT.

$$240^2 = 57600$$
$$12^2 = 144$$

57744
57600
—————
144

480)1440(·3
1440

∴ UT = 240·3

Fig. 25.

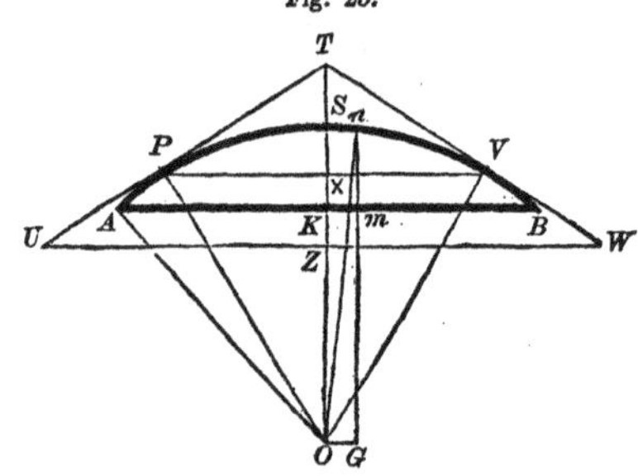

ZT : TU : : PT : TO, that is,

12 : 240·3 : : 146 : 2923·65

TO = 2923·65
SO = 2920·
—————
3·65 = TS.

SK = 8
ST = 3·65
—————
11·65 = TK.

Consequently the points P, S, V, K, A, B are known, and may be staked out on the field. These points, as well as the lengths of the ordinates, may be found in the office before going on the ground.

Let the points and ordinates be found right and left of the middle point K.

Let Km = OG = 8

$On^2 = 2920^2 = 8526400\cdot$

$OG^2 = 8^2 = 64\cdot$

```
                 8526336(2919·
  291₂ =         84681
                 -----
  5829             58236
                   52461
                   -----
  58389           577500(2919·98
                  525501
                  ------
  58398           5199900
                  -------

          2919·98
          2912·00
          -------
             7·98  the first ordinate
```

right or left of K.

The square root of 8526336 cannot be approached in a direct manner from a table up to the square of 1600. In finding 2919·98, the square of 291 was commenced with, and the extraction of the root by the common method continued.

By taking the fourth part of 8526336 = 2131584, the root is more easily found by the table thus:—

```
            2131584
  1459₂ =   2128681
            -------
  2918)       29030(·994
              26262
              -----
               27680
               26262
               -----
                14180
                11672
                -----

          1459·994
                 2
          --------
          2919·988  From
          2912      take
          --------
             7·989  ordinate.
```

To show the ease by which these ordinates may be found when the plan is established, I will give the figures employed in finding one or two more.

$$
\begin{array}{rrl}
2920^2 & = & 8526400 \\
16^2 & = & 256 \\
 & & 4)8526144 \\
 & & 2131536 \\
1459^2 & = & 2128681 \\
2918 & & 28550(\cdot 944 \\
 & & 26262 \\
 & & \cdot 12180 \\
 & & 11672 \\
 & & 12080 \\
 & & 11672 \\
\end{array}
$$

$$
\begin{array}{r}
1459\cdot 944 \\
2 \\
\hline
2919\cdot 888 \\
2912\cdot \\
\hline
7\cdot 888
\end{array}
$$

7·888 the second ordinate right and left of K

Required the tenth ordinate right and left of K.

$$
\begin{array}{rrrl}
 & 2920^2 & = & 8526400 \\
10 \times 8 = & 80^2 & = & 6400 \\
 & & 4) & 8520000 \\
 & & & 2130000 \\
 & 1459^2 & = & 2128681 \\
 & 2918) & & 13190 \quad (\cdot 452 \\
 & & & 11672 \\
 & & & 15180 \\
 & & & 14590 \\
 & & & 5900 \\
 & & & 5836 \\
\end{array}
$$

$$
\begin{array}{r}
1459\cdot 452 \\
2 \\
\hline
2918\cdot 904 \\
2912\cdot \\
\hline
6\cdot 904
\end{array}
$$

6·904 = 10th ordinate to the right and left of K.

COMPOUND CURVES.

Compound railroad curves, A, B, C, D, E, F, G, Fig. 26, composed of straight lines and circular arcs, have common normals, OH, OP, PI, QJ, KR, and therefore common tangents where the

Fig. 26.

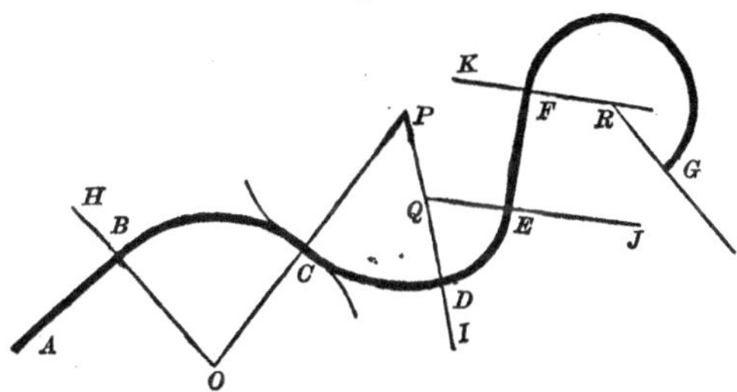

arcs are joined. The normals are perpendicular to the straight portions of the road also; OH is perpendicular to AB, EF is perpendicular to QJ and KR.

This kind of curve is adopted where the railroad is required to pass through given points, as C, D, E, F, or to avoid obstructions.

PROPOSITION X.

To find the Radii OB, CQ, to connect two straight lines of Railroad, AB, DE; the Road has to pass from the point B, through the point C, and to touch the straight road EF at any point D.

Fig. 27.

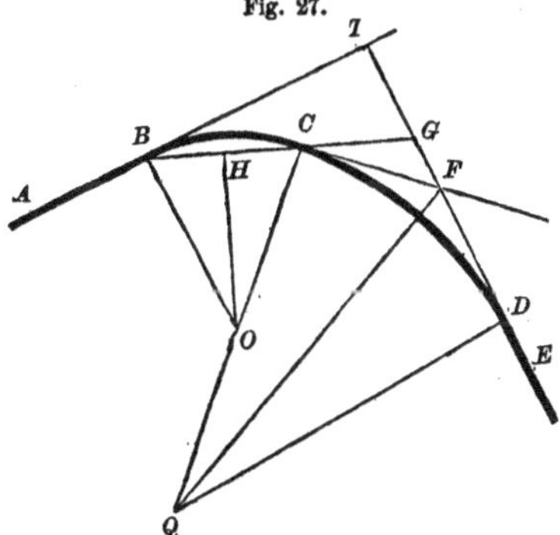

GEOMETRICAL CONSTRUCTION.

Join B and C, make the angle BCO = OBC, which is supposed

to be given, $= 90^\circ -$ TBC. Draw BO perpendicular to AB, then OB $=$ CO.

With OB as radius, describe the arc BC; draw CF perpendicular to CQ, and produce DE to meet it in F, make DF $=$ CF, and draw DQ, perpendicular to EF, to meet CQ in Q,

$$CQ = QD.$$

Hence the radii OB, QD, are determined.

EXAMPLE.

Let BT $= 1100$ feet, BC $= 700$ feet; the angle BTD $= 92^\circ$; and TBG $= 24^\circ$.

$$180^\circ - 92^\circ - 24^\circ = 64^\circ = \text{TGB}.$$

$$\text{sin. TGB} : \text{BT} :: \text{sin. BTG} : \text{BG}.$$

$$\text{sin. } 92^\circ = \text{sin. } 88^\circ.$$

8988 : 1100 : : ·9994
1100
999400
999400

8988) 1099·3400 (1223·12 = BG.
8988
20054
17976
20780
17976
28040
26964
10760
8988
17720
17976

BG = 1223·12
BC = 700·
CG = 523·12

Angle CGF = 92° + 24° = 116°.
" GCF = 24°.
" CFG = 180° — 116° —24° = 40°
Sin. CFG : CG : : sin. CGF : CF.

```
·6428 : 523·12 : : ·8988
                   523·12
                   ------
                   17976
                   8988
                 26964
                17976
               44940
               ----------
·6428          470·180256 (731.45
               44996
               -----
                20220
                19284
                -----
731·45 = CF      9362
                 6428
                 ----
                 29345
                 25712
                 -----
                  36336
                  32140
                  -----
```

Angle CFD = 180° — 40° = 140°.
half of CFD = CFQ = 70°.
CQF = 99° — 70° = 20°.
Sin. CQF : CF : : sin. CFQ : CQ.

```
·342 : 731·45 : : ·94
           ·94
        ------
        292580
       658305
       -------
·342)  687·5630 (2010·4 = CQ.
       684
       ---
         356
         342
         ---
          1430
          1368
          ----
```

Sin. BOH : BH : : sin. BHO : OB.
BOH = TBG = 24°.

```
·407  :  350  : :  1·000
               1
         ─────────
·407       3500  (859·95 = OB.
           3256
           ────
            2440
            2035
            ────
             4050
             3663
             ────
              3870
              3663
              ────
               2070
               2035
               ────
```

The radii required have been found by the mere application of a table of natural sines. OB = 859·95 feet, and QC = 2010·4 feet.

PROBLEM.

To lay out a Railroad curve by means of ordinates or offsets in rational numbers.

Fig. 28.

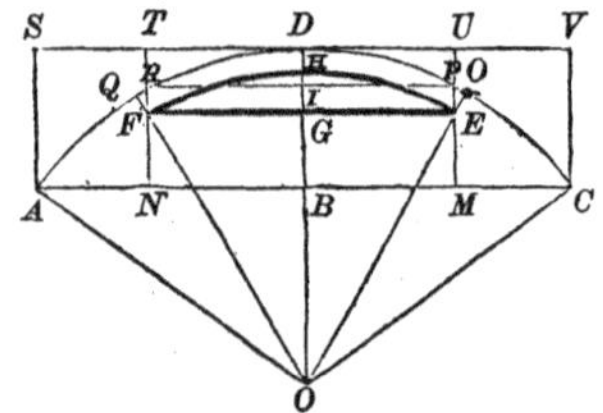

Let $OC = \frac{n}{2} + \frac{\left(\frac{a}{2}\right)^2}{\frac{n}{2}}$; BC = a; BM =

GE = IP = b; $OB = \frac{n}{2} - \frac{\left(\frac{a}{2}\right)^2}{\frac{n}{2}}$;

$$OB^2 + BC^2 = OC_2$$

$$\left\{\frac{n}{2} - \frac{\left(\frac{a}{2}\right)^2}{\frac{n}{2}}\right\}^2 + a^2 = \left\{\frac{n}{2} + \frac{\left(\frac{a}{2}\right)^2}{\frac{n}{2}}\right\}^2$$

When each side of this equation is developed it becomes,

$$\left(\frac{n}{2}\right)^2 + 2\left(\frac{a}{2}\right)^2 + \frac{\left(\frac{a}{2}\right)^4}{\frac{n^2}{2}} = \left(\frac{n}{2}\right)^2 + 2\left(\frac{a}{2}\right)^2 + \frac{\left(\frac{a}{2}\right)^4}{\left(\frac{n}{2}\right)^2},$$

or identical; consequently OBC is rational whatever values be given to n and a,

$$\text{Let } \frac{n}{2} = 653, \text{ then, if } a = 50,$$

$$OC = 653\frac{625}{653} = OD.$$

$$OB = 653 - \frac{625}{653}.$$

$$DB = \frac{1250}{653} = \frac{2\left(\frac{a}{2}\right)^2}{\frac{n}{2}}$$

I have now established that when the radius OC =

$$\frac{n}{2} \times \frac{\left(\frac{a}{2}\right)^2}{\frac{n}{2}}$$

and BC = a, then CV = MU = BD = NT = AS =

$$\frac{\frac{a^2}{2}}{\frac{n}{2}}.$$

By the same reasoning, if BM = BN = FG = GE = PI = IR = ID = DU be put = b, then OE = OH = OF =

$$\frac{n}{2} + \frac{\left(\frac{b}{2}\right)^2}{\frac{n}{2}} \text{ and } OG = \frac{n}{2} - \frac{\left(\frac{b}{2}\right)^2}{\frac{n}{2}}$$

$$\text{Also } GH = ID = \frac{\frac{b^2}{2}}{\frac{n}{2}} = PU = RT.$$

If, for example, $BM = DU = 40 = b$, $PU = \frac{800}{653}$, taking $\frac{n}{2} = 653$, as in the last case.

The way in which the following table is arranged becomes evident; for, if the radius $= 648\frac{625}{648}$, the ordinates outside the curve at 50, and at 20, becomes

$$\frac{1250}{648} \text{ and } \frac{200}{648} \text{ respectively.}$$

In this case, $\frac{n}{2} = 648$; $a = 50$, or $AC = 100$ feet $= 2a$; $b = 20$.

It must be observed that the segment FGEH very nearly coincides with the segment RIPD, when the point H covers the point D, and the point G the point I.

Or, in other words, F will nearly fall upon R, and E upon P. At first sight it might appear that these segments of circles would exactly coincide, but this is not the case; for, as before observed, in moving GH to coincide with JD, the point F will not exactly coincide with R, nor will E with P.

If this should happen, $FR = EP = DH$. But $DH = FQ$; and $FR = FQ$, which is absurd.

But in the case before us, FR is extremely near DH or FQ. If required, the difference can be readily found, as the right angled triangles FQR and OGF are similar,

$$GO : OF :: OF : FR.$$

$$\text{that is } \frac{n}{2} - \frac{\frac{b^2}{2}}{\frac{n}{2}} : \frac{n}{2} + \frac{\frac{b^2}{2}}{\frac{n}{2}} :: QF : FR.$$

If $\frac{n}{2} = 2000$, and $b = 40$, then the above proportion becomes

$$2000 - \frac{800}{2000} : 2000 + \frac{800}{2000} :: QF : FR.$$

$$1999\tfrac{3}{5} : 2000\tfrac{2}{5} :: QF : FR.$$

ORDINATES.

Radii.	50	40	30	20	10	Reciprocals.
600 $\frac{625}{600}$	$\frac{1250}{600}$	$\frac{800}{600}$	$\frac{450}{600}$	$\frac{200}{600}$	$\frac{50}{600}$	·0016666
601 $\frac{625}{601}$	$\frac{1250}{601}$	$\frac{800}{601}$	$\frac{450}{601}$	$\frac{200}{601}$	$\frac{50}{601}$	·0016639
602 $\frac{625}{602}$	$\frac{1250}{602}$	$\frac{800}{602}$	$\frac{450}{602}$	$\frac{200}{602}$	$\frac{50}{602}$	·0016611
603 $\frac{625}{603}$	$\frac{1250}{603}$	$\frac{800}{603}$	$\frac{450}{603}$	$\frac{200}{603}$	$\frac{50}{603}$	·0016584
604 $\frac{625}{604}$	$\frac{1250}{604}$	$\frac{800}{604}$	$\frac{450}{604}$	$\frac{200}{604}$	$\frac{50}{604}$	·0016556
605 $\frac{625}{605}$	$\frac{1250}{605}$	$\frac{800}{605}$	$\frac{450}{605}$	$\frac{200}{605}$	$\frac{50}{605}$	·0016529
606 $\frac{625}{606}$	$\frac{1250}{606}$	$\frac{800}{606}$	$\frac{450}{606}$	$\frac{200}{606}$	$\frac{50}{606}$	·0016502
607 $\frac{625}{607}$	$\frac{1250}{607}$	$\frac{800}{607}$	$\frac{450}{607}$	$\frac{200}{607}$	$\frac{50}{607}$	·0016474
608 $\frac{625}{608}$	$\frac{1250}{608}$	$\frac{800}{608}$	$\frac{450}{608}$	$\frac{200}{608}$	$\frac{50}{608}$	·0016447
609 $\frac{625}{609}$	$\frac{1250}{609}$	$\frac{800}{609}$	$\frac{450}{609}$	$\frac{200}{609}$	$\frac{50}{609}$	·0016420
610 $\frac{625}{610}$	$\frac{1250}{610}$	$\frac{800}{610}$	$\frac{450}{610}$	$\frac{200}{610}$	$\frac{50}{610}$	·0016393
611 $\frac{625}{611}$	$\frac{1250}{611}$	$\frac{800}{611}$	$\frac{450}{611}$	$\frac{200}{611}$	$\frac{50}{611}$	·0016367
612 $\frac{625}{612}$	$\frac{1250}{612}$	$\frac{800}{612}$	$\frac{450}{612}$	$\frac{200}{612}$	$\frac{50}{612}$	·0016340
613 $\frac{625}{613}$	$\frac{1250}{613}$	$\frac{800}{613}$	$\frac{450}{613}$	$\frac{200}{613}$	$\frac{50}{613}$	·0016313
614 $\frac{625}{614}$	$\frac{1250}{614}$	$\frac{800}{614}$	$\frac{450}{614}$	$\frac{200}{614}$	$\frac{50}{614}$	·0016287
615 $\frac{625}{615}$	$\frac{1250}{615}$	$\frac{800}{615}$	$\frac{450}{615}$	$\frac{200}{615}$	$\frac{50}{615}$	·0016260
616 $\frac{625}{616}$	$\frac{1250}{616}$	$\frac{800}{616}$	$\frac{450}{616}$	$\frac{200}{616}$	$\frac{50}{616}$	·0016234
617 $\frac{625}{617}$	$\frac{1250}{617}$	$\frac{800}{617}$	$\frac{450}{617}$	$\frac{200}{617}$	$\frac{50}{617}$	·0016207
618 $\frac{625}{618}$	$\frac{1250}{618}$	$\frac{800}{618}$	$\frac{450}{618}$	$\frac{200}{618}$	$\frac{50}{618}$	·0016181
619 $\frac{625}{619}$	$\frac{1250}{619}$	$\frac{800}{619}$	$\frac{450}{619}$	$\frac{200}{619}$	$\frac{50}{619}$	·0016155
620 $\frac{625}{620}$	$\frac{1250}{620}$	$\frac{800}{620}$	$\frac{450}{620}$	$\frac{200}{620}$	$\frac{50}{620}$	·0016129

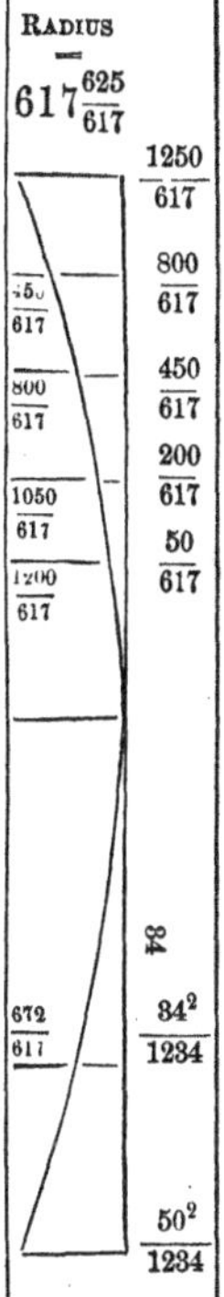

Radius = $617\frac{625}{617}$

$\frac{450}{617}$ $\frac{800}{617}$ $\frac{1050}{617}$ $\frac{1200}{617}$ $\frac{672}{617}$

$\frac{1250}{617}$ $\frac{800}{617}$ $\frac{450}{617}$ $\frac{200}{617}$ $\frac{50}{617}$

84

$\frac{84^2}{1234}$

$\frac{50^2}{1234}$

ORDINATES.

Radii.	50	40	30	20	10	Reciprocals.
621 625/621	1250/621	800/621	450/621	200/621	50/621	·0016103
622 625/622	1250/622	800/622	450/622	200/622	50/622	·0016077
623 625/623	1250/623	800/623	450/623	200/623	50/623	·0016050
624 625/624	1250/624	800/624	450/624	200/624	50/624	·0016025
625 625/625	1250/625	800/625	450/625	200/625	50/625	·0016000
626 625/626	1250/626	800/626	450/626	200/626	50/626	·0015974
627 625/627	1250/627	800/627	450/627	200/627	50/627	·0015948
628 625/628	1250/628	800/628	450/628	200/628	50/628	·0015924
629 625/629	1250/629	800/629	450/629	200/629	50/629	·0015898
630 625/630	1250/630	800/630	450/630	200/630	50/630	·0015873
631 625/631	1250/631	800/631	450/631	200/631	50/631	·0015848
632 625/632	1250/632	800/632	450/632	200/632	50/632	·0015823
633 625/633	1250/633	800/633	450/633	200/633	50/633	·0015798
634 625/634	1250/634	800/634	450/634	200/634	50/634	·0015773
635 625/635	1250/635	800/635	450/635	200/635	50/635	·0015748
636 625/636	1250/636	800/636	450/636	200/636	50/636	·0015723
637 625/637	1250/637	800/637	450/637	200/637	50/637	·0015699
638 625/638	1250/638	800/638	450/638	200/638	50/638	·0015674
639 625/639	1250/639	800/639	450/639	200/639	50/639	·0015649
640 625/640	1250/640	800/640	450/640	200/640	50/640	·0015625
641 625/641	1250/641	800/641	450/641	200/641	50/641	·0015601
642 625/642	1250/642	800/642	450/642	200/642	50/642	·0015576
643 625/643	1250/643	800/643	450/643	200/643	50/643	·0015552
644 625/644	1250/644	800/644	450/644	200/644	50/644	·0015528
645 625/645	1250/645	800/645	450/645	200/645	50/645	·0015504
646 625/646	1250/646	800/646	450/646	200/646	50/646	·0015480

Radius = 637 625/637

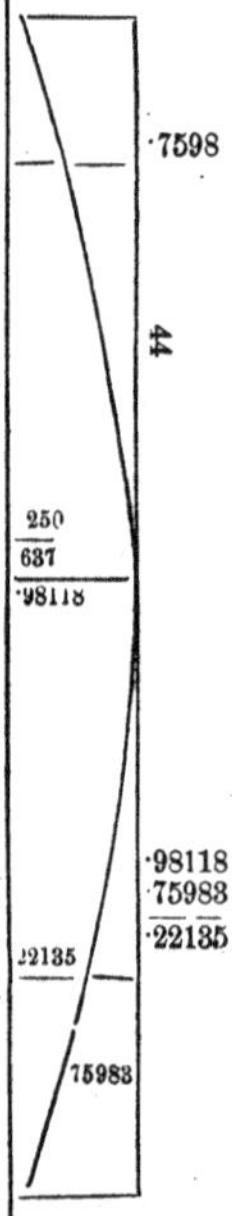

·98118
·75983
·22135

·0015699
22
0031398
0031398
·0345378
22
0690756
690756
·7598316

ORDINATES.

Radii.	50	40	30	20	10	Reciprocals.
$647\frac{625}{647}$	$\frac{1250}{647}$	$\frac{800}{647}$	$\frac{450}{647}$	$\frac{200}{647}$	$\frac{50}{647}$	·0015456
$648\frac{625}{648}$	$\frac{1250}{648}$	$\frac{800}{648}$	$\frac{450}{648}$	$\frac{200}{648}$	$\frac{50}{648}$	·0015432
$649\frac{625}{649}$	$\frac{1250}{649}$	$\frac{800}{649}$	$\frac{450}{649}$	$\frac{200}{649}$	$\frac{50}{649}$	·0015408
$650\frac{625}{650}$	$\frac{1250}{650}$	$\frac{800}{650}$	$\frac{450}{650}$	$\frac{200}{650}$	$\frac{50}{650}$	·0015385
$651\frac{625}{651}$	$\frac{1250}{651}$	$\frac{800}{651}$	$\frac{450}{651}$	$\frac{200}{651}$	$\frac{50}{651}$	·0015361
$652\frac{625}{652}$	$\frac{1250}{652}$	$\frac{800}{652}$	$\frac{450}{652}$	$\frac{200}{652}$	$\frac{50}{652}$	·0015337
$653\frac{625}{653}$	$\frac{1250}{653}$	$\frac{800}{653}$	$\frac{450}{653}$	$\frac{200}{653}$	$\frac{50}{653}$	·0015314
$654\frac{625}{654}$	$\frac{1250}{654}$	$\frac{800}{654}$	$\frac{450}{654}$	$\frac{200}{654}$	$\frac{50}{654}$	·0015291
$655\frac{625}{655}$	$\frac{1250}{655}$	$\frac{800}{655}$	$\frac{450}{655}$	$\frac{200}{655}$	$\frac{50}{655}$	·0015267
$656\frac{625}{656}$	$\frac{1250}{656}$	$\frac{800}{656}$	$\frac{450}{656}$	$\frac{200}{656}$	$\frac{50}{656}$	·0015244
$657\frac{625}{657}$	$\frac{1250}{657}$	$\frac{800}{657}$	$\frac{450}{657}$	$\frac{200}{657}$	$\frac{50}{657}$	·0015221
$658\frac{625}{658}$	$\frac{1250}{658}$	$\frac{800}{658}$	$\frac{450}{658}$	$\frac{200}{658}$	$\frac{50}{658}$	·0015198
$659\frac{625}{659}$	$\frac{1250}{659}$	$\frac{800}{659}$	$\frac{450}{659}$	$\frac{200}{659}$	$\frac{50}{659}$	·0015175
$660\frac{625}{660}$	$\frac{1250}{660}$	$\frac{800}{660}$	$\frac{450}{660}$	$\frac{200}{660}$	$\frac{50}{660}$	·0015152
$661\frac{625}{661}$	$\frac{1250}{661}$	$\frac{800}{661}$	$\frac{450}{661}$	$\frac{200}{661}$	$\frac{50}{661}$	·0015129
$662\frac{625}{662}$	$\frac{1250}{662}$	$\frac{800}{662}$	$\frac{450}{662}$	$\frac{200}{662}$	$\frac{50}{662}$	·0015106
$663\frac{625}{663}$	$\frac{1250}{663}$	$\frac{800}{663}$	$\frac{450}{663}$	$\frac{200}{663}$	$\frac{50}{663}$	·0015083
$664\frac{625}{664}$	$\frac{1250}{664}$	$\frac{800}{664}$	$\frac{450}{664}$	$\frac{200}{664}$	$\frac{50}{664}$	·0015060
$665\frac{625}{665}$	$\frac{1250}{665}$	$\frac{800}{665}$	$\frac{450}{665}$	$\frac{200}{665}$	$\frac{50}{665}$	·0015038
$666\frac{625}{666}$	$\frac{1250}{666}$	$\frac{800}{666}$	$\frac{450}{666}$	$\frac{200}{666}$	$\frac{50}{666}$	·0015015
$667\frac{625}{667}$	$\frac{1250}{667}$	$\frac{800}{667}$	$\frac{450}{667}$	$\frac{200}{667}$	$\frac{50}{667}$	·0014993
$668\frac{625}{668}$	$\frac{1250}{668}$	$\frac{800}{668}$	$\frac{450}{668}$	$\frac{200}{668}$	$\frac{50}{668}$	·0014970
$669\frac{625}{669}$	$\frac{1250}{669}$	$\frac{800}{669}$	$\frac{450}{669}$	$\frac{200}{669}$	$\frac{50}{669}$	·0014948
$670\frac{625}{670}$	$\frac{1250}{670}$	$\frac{800}{670}$	$\frac{450}{670}$	$\frac{200}{670}$	$\frac{50}{670}$	·0014925

Radius = $657\frac{625}{657}$

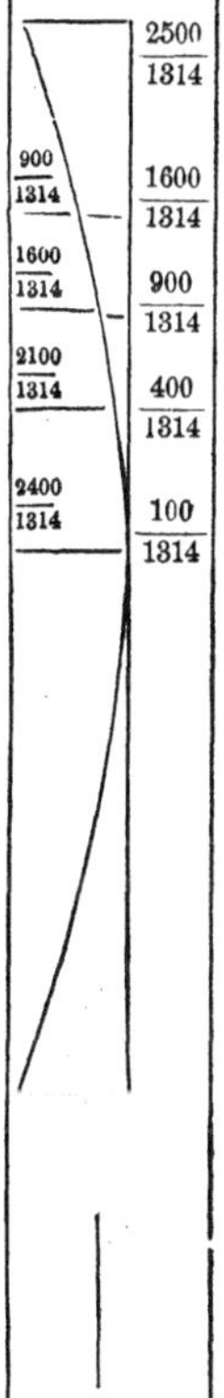

$\frac{2500}{1314}$ $\frac{900}{1314}$ $\frac{1600}{1314}$ $\frac{1600}{1314}$ $\frac{900}{1314}$ $\frac{2100}{1314}$ $\frac{400}{1314}$ $\frac{2400}{1314}$ $\frac{100}{1314}$

Hence the near approach of QF to FR, in this and other similar cases, is easily observed.

The middle ordinate, on the concave side of the curve, or its equal, either of the extreme ordinates on the convex side of the curve is found mathematically exact. By what I have established, the ordinates for any radius or for any chord may be set down without labor, as an example or two will show.

EXAMPLE.

For a radius between 6973 and 6975 feet, and a long chord of 300 feet, or 3 chains, what is the ordinate in the middle at the concave side of the curve?

$$\text{Take the radius} = 6973\,\frac{5625}{6973}$$

$$\frac{300}{2} = 150;\ \left(\frac{150}{2}\right)^2 = 5625$$

The square of 150 may be found from a table of squares, &c.

$$\text{The required ordinate} = \frac{11250}{6973}\ \text{feet exactly.}$$

EXAMPLE.

For a radius between 68975 and 68977 feet, and a chord of 400 feet, what is the length of the ordinates in the middle at the concave side of the curve, or at the extreme ends on the convex side.

$$\frac{400}{4} = 100$$

$$\text{Put the radius} = 68975\,\frac{10000}{68975}$$

$$\text{The length of the required ordinates will be} = \frac{20000}{68975}$$

EXAMPLE.

If the radius of a curve $= 3476\,\frac{625}{3476}$ feet; chord = 100 feet; required the ordinate in middle, and 26 feet from the middle both on the concave and convex sides.

$$\text{Ordinate in the middle} = \frac{1250}{3476},\ 26^2 = 676$$

$$\text{Ordinate on the convex side 26ft. from the middle} = \frac{338}{3476}$$

Ordinate on the concave side 26ft. from the middle =

$$\frac{1250}{3476} - \frac{338}{3476} = \frac{912}{3476}$$

It is clear that the rule will apply to all circular curves, without mental labor. Any number, as 130, may be reduced to the form.

$$x + \frac{625}{x} = 130$$

by solving this equation, or

$$x^2 - 130x = -625$$

$$\therefore x = 125$$

$$125 \frac{625}{125} = 130$$

In a 1° curve, that is a curve in which a chord of 100 feet subtends 1° at the centre, the radius = 5729·65 feet.

2° curve, radius =	2864·93	"
3° curve, radius =	1910·07	"
4° curve, radius =	1432·70	"
5° curve, radius =	1146·27	"
6° curve, radius =	955·36	"
7° curve, radius =	819·01	"
8° curve, radius =	716·77	"
9° curve, radius =	637·28	"
10° curve, radius =	573·66	"

Any of these numbers is readily found by dividing fifty by the natural sine of the given degrees. Thus natural sine of 5° = ·08716, and

$$\frac{50}{\cdot 08716} = 573\cdot 66 \text{ feet.}$$

Without solving an equation, these radii or others may be reduced to the form $x \times \frac{625}{x}$, by taking the nearest whole number to x and dividing it into 625.

$$\frac{625}{572} = 1\cdot 09$$

From 573·66
Take 1·09
572·57

$$\text{Then } 572\cdot 57 \frac{625}{572\cdot 57} = 573\cdot 66$$

EXAMPLE.

In a 2° curve, the radius = 2864·93 feet, what is the length of the middle ordinate and the length of an ordinate 14 feet from the middle on both the concave and convex sides of the curve; as usual, measuring from the tangent at the middle point and from the chord?

$$\frac{625}{2864} = \cdot 22$$

$$\begin{array}{r} 2864{\cdot}93 \\ {\cdot}22 \\ \hline 2864{\cdot}71 \end{array}$$

$$\therefore\ 2864{\cdot}71 \frac{625}{2864{\cdot}71} = \text{Radius.}$$

$$\text{Middle ordinate} = \frac{1250}{2864{\cdot}71}.$$

$$14^2 = 196$$

$$\text{Ordinate on the convex side 14 feet from the middle} = \frac{98}{2864{\cdot}71}$$

$$\begin{array}{r} 1250 \\ 98 \\ \hline 1152 \end{array}$$

$$\text{Ordinate on the concave side 14 feet from the middle} = \frac{1152}{2864{\cdot}71}.$$

If the abscissa = 15 feet, then

$$\text{Middle ordinate} = \frac{2500}{2864{\cdot}71 \times 2}, \text{ as before.}$$

$$\text{Ordinate concave} = \frac{2275}{5729{\cdot}42}.$$

$$\text{Ordinate convex} = \frac{225}{5729{\cdot}42}.$$

PROPOSITION XI.

One of the two Radii of the Compound Curve, and its starting and closing points being given, to find the other Radius.

Let AB, DE, Figs. 29, 30, be the tangents, B and D the starting and closing points of the curve. Draw to the tangents the perpen-

Fig. 29.

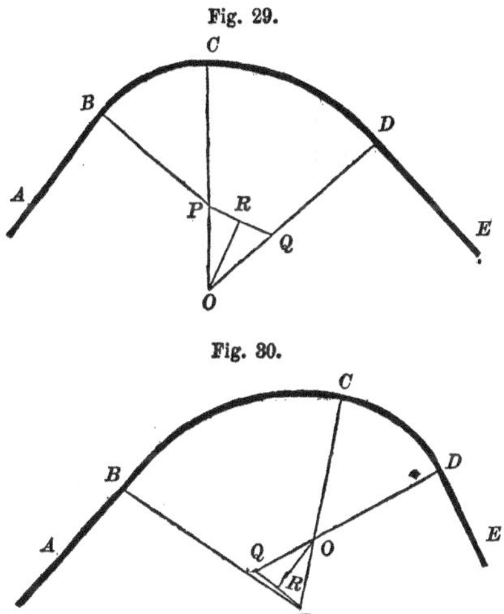

Fig. 30.

diculars BP = DQ = the given radius; join P and Q and bisect it in R; draw RO perpendicular to PQ, meeting DQ or DQ produced in O; join OP and prolong it if necessary till CP = PB; then the points P, O, are the centrès of the arcs BC and CD, that constitute the compound curve. OC is the required radius. Because PC and QD were made equal to PB, OC = OD, since OP = OQ.

Fig. 31.

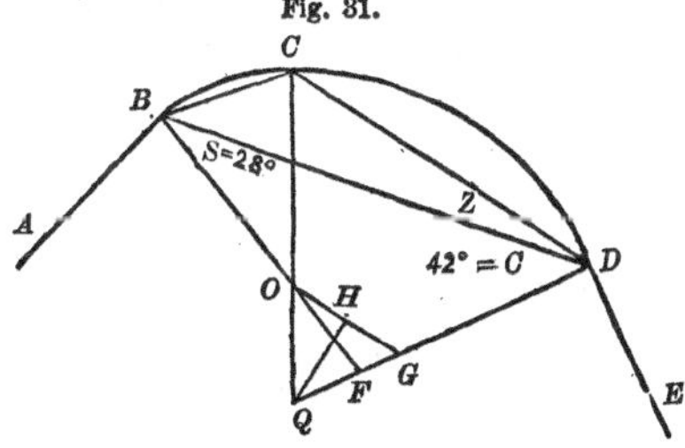

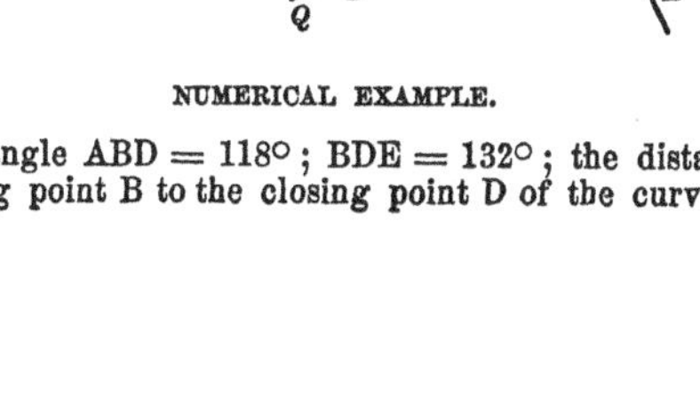

NUMERICAL EXAMPLE.

Let the angle ABD = 118°; BDE = 132°; the distance from the starting point B to the closing point D of the curve = 2150 feet; the given radius, OB = OC = GD = 975 feet. Required QD = QC.

Angle DBF = 118° — 90° = 28°.
BDF = 132 — 90 = 42°.
BFD = 180° — 42° — 28° = 110°.

sin. 110° : 2150 :: sin. 42° : BF = 1698·1 feet.

Log. sin. 42°	9·825511
log. 2150	3·332438
	13·157949
Log. sin. 110° = log. sin. 70° ..	9·927986
1698·1 = BF, log.	3·229963

1698·1
975·
723·1 = OF.

Sin. 110° : 2150 :: sin. 28° : FD =

Log. sin 28°	9·671609
log. 2150	3·332438
	13·004047
Log. sin. 110°	9·927986
1191·4 = FD, log......... ...	3·076061

1191.4
975·
216·4 = FG.

In the triangle OFG, the two sides OF, FG, and the included angle OFG, are known; hence, by trigonometry, the remaining parts of the triangle may be found.

As. 723·1 + 216·4 : 723·1 — 216·4 :: tan. ½ (180 —110) : the tangent of half the difference of the angles OGF, FOG. That is,

9395 : 506·7 :: tan. 35° : tan.

Log. tan. 35°	9·845227
log. 506·7	2·704751
	12·149978
log. 939·5	2·972897
log. tan. 8° .. 33′	9·177081

35° = half sum of angles FGO and FOG.
8 33′ = half difference of angles FGO and FOG.
43° 33′ = angle OGQ.

35° 0′
8 33
26° 27′ = angle FOG.

Sin. 43° 33′ : OF :: sin. 110° : OG = 986·21 feet.

Log. sin. 110°	9·972986
log. OF = 723·1	2·859198
	12·832184
log. sin. 43° 33′	9·838211
OG = 986·21, log.	2·993973

2) 986·21
493·105 = HG.

90° 0′
43 33
46 27 = angle HQG.

Sin. 46° 27′ : HG :: sin. 90 : QG = 680·36

Log. sin. 90°	10·000000
log. 493·105	2·692939
	12·692939
log. sin. 46° 27′	9·860202
QG = 680·36 log.	2·832737

975 = GD.
680·36 = QG.
1655·36 = QD the radius required.

Fig. 32.

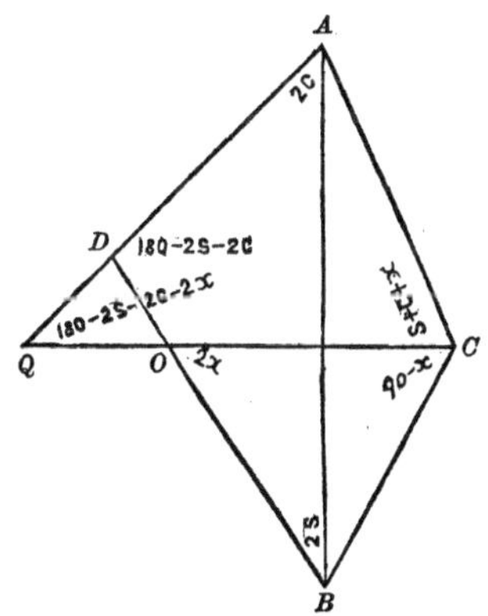

Let the three points Q, O, C, be in a straight line, and QA = QC as well as OC = OB, then the angle ACB = 90° + half DAB + half DBA. See Fig. 32.

$$\text{Put } 2x = \text{angle COB.}$$
$$2c = \text{angle DAB.}$$
$$2s = \text{angle DBA.}$$

$$\text{Then, } 180^\circ - 2s - 2c = \text{angle ADB}$$
$$180 - 2s - 2c - 2x = \text{AQC.}$$

$$\left.\begin{array}{r}\therefore s + c + x = \text{QCA.} \\ 90^\circ - x = \text{OCB.}\end{array}\right\} \text{add.}$$

$\therefore s + c + 90^\circ$ = ACB, which was to be proved. I shall now find the relation existing between the radii AQ and OB, and the angle of the triangle CAB, next the greater radius AQ, that is CAB. Fig. 33.

Fig. 33.

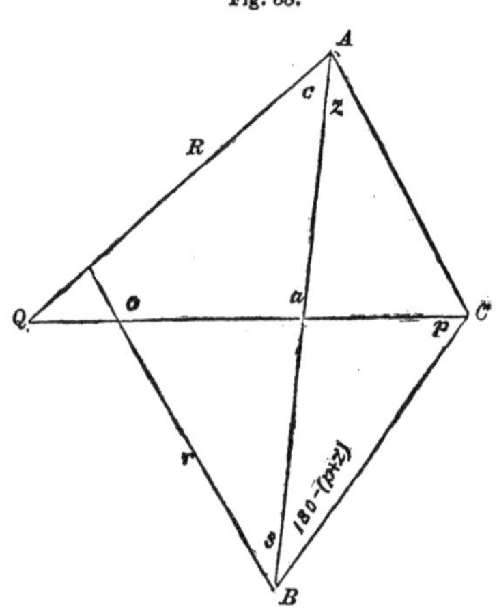

Put a = AB, the given distance between the starting and closing points of the curve.

Put p = angle ACB,
c = " QAB,
s = " OBA,

z, the required angle, = BAC. R = QA, and r = OB. Angle OBC = $180^\circ - (p + z) + s = 180^\circ - (p - s + z)$.

$$\sin. p : a :: \sin. z : \frac{a \sin. z}{\sin. p} = \text{BC.}$$

BC = 2r cos. (OBC) = 2r cos. (180° — (p — s + z) = 2r cos. . (p — s + z)

$$\therefore \frac{a \sin. z}{\sin. p} = 2r \cos. (p - s + z)$$

$$= 2r \left\{ \cos. (p - s) \cos. z - \sin. (p - s) \sin. z \right\}$$

Dividing both sides of the equation by sin. z,

$$\frac{a}{\sin. p} = 2r \left\{ \cos. (p - s) \cot. z - \sin. (p - s) \right\}.$$

$$\therefore \frac{a}{2r \sin. p \cos. (p - s)} + \tan. (p - s) = \cot. z,$$

the relation existing between r and z.

Again, sin. p : a : : sin. (180° — ($p + z$) : AC,
also, AC = 2R cos. ($c + z$).

$$\therefore 2R \cos. (c + z) = \frac{a \sin. (p + z)}{\sin. p},$$

$$2R \left\{ \cos. c \cos. z - \sin. c \sin. z \right\} = \frac{a \left\{ \sin. p \cos. z - \cos. p \sin. z \right\}}{\sin. p.}$$

Dividing both sides by sin. z, then,

$$2R \left\{ \cos. c \cot. z - \sin. c \right\} = \frac{a \left\{ \sin. p \cot. z - \cos. p \right\}}{\sin. p.}$$

$$2R \left\{ \cos. c \cot. z - \sin. c \right\} = a \left\{ \cot. z - \cot. p \right\}$$

$$2R \cos. c \cot. z - 2R \sin. c = a \cot. z - a \cot. p$$

$$\cot. z = \frac{2R \sin. c - a \cot. p}{2R \cos. c - a},$$

which determines z without involving r.

Let A be the starting, and B the closing point of the curve ACB, compounded of AC and CB; the angles EAB or FAB and IBA or ABH, can generally be taken in the field.

OB and QA are respectively perpendicular to HI and EF.

Put s = angle OBA;
c = angle QAB;
a = AB;
R = AQ;
r = OB;

∴ AG = R cos. c; BL = r cos. s;
GL = OP = R cos. c + r cos. s — a;
QG = R sin. c; OL = PG = r sin. s;
PQ = R sin. c — r sin. s.

Fig. 84.

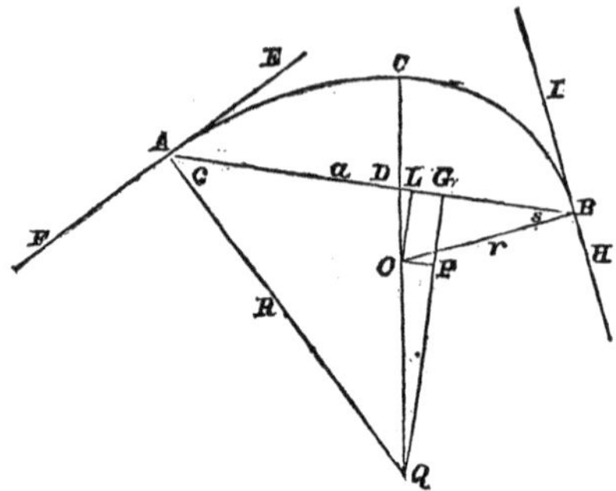

But $OQ^2 = (R - r)^2 = OP^2 + PQ^2$.

This is general, as the three sides of the right angled triangle QPQ are composed of differences, for it is well known that $(R - r)^2 = (r - R)^2$ and so with other differences, as $(R \sin c - r \sin. s)^2 = (r \sin. s - R \sin. c)^2$.

It must be remembered also that, $\sin.^2 + \cos.^2 = 1$; and that $\cos. (c + s) = \cos. s \cos. c - \sin. s \sin. c$.

Now, to square the sides of the triangle OPQ;—

$(R \cos. c + r \cos. s - a)^2 = OP^2$;

Fig. 85.

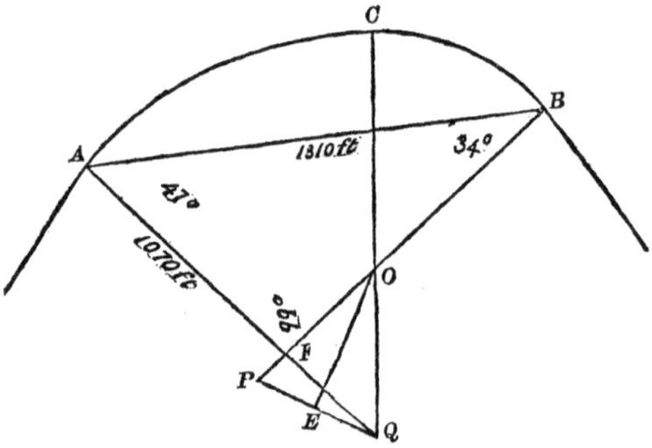

$$OP^2 = R^2 \cos.{}^2 c + 2\, r\, R \cos.\, c \cos.\, s + r^2 \cos.{}^2 s - 2\, a\, r \cos.\, s + a^2 - 2\, a\, R \cos.\, c.$$

$$PQ^2 = R^2 \sin.{}^2 c - 2\, r\, R \sin.\, c \sin.\, s + r^2 \sin.{}^2 s.$$

$$OP^2 + PQ^2 = R^2 + 2\, r\, R \operatorname{Cos.} (c + s) + r^2 - 2\, a\, r \cos.\, s + a^2 - 2\, a\, R \cos.\, c = OQ^2 = R^2 - 2\, r\, R + r^2$$

$$\therefore 2\, r\, R \cos.\, (c + s) - 2\, a\, r \cos.\, s + 2\, r\, R = 2\, a\, R \cos.\, c - a^2.$$

$$r = \frac{2\, a\, R \cos c - a^2}{2\, R \cos.\, (c + s) - 2\, a \cos.\, s + 2\, R}$$

$$r = \frac{a\, (2\, R \cos.\, c - a)}{2 \left\{ R \cos.\, (c + s) - a \cos.\, s + R \right\}}$$

EXAMPLE.

Given AB = 1310 feet, radius AQ = 1070 feet, required radius OB, angle QAB = 47°; OBA = 34°.

Sin. 99° : 1310 :: sin. 34° : AF = 741·7

Log. sin. 34°	9·747562
log. 1310	3·117271
	12·864833
log. sin. 81° = log. sin. 99° ..	9·994620
log. 741·7	2·870213

1070·
741·7
328·3 = QF.

Sin. 99° : 1310 :: sin. 47° : FB = 970·

Log. sin. 47°	9·864127
log. 1310	3·117271
	12·981398
log. sin. 99°	9·994620
log. 970·	2·986778

1070
970
100 = FD.

In the triangle QFD we have the two sides QF, FD, and the contained angle DFQ, given, to find the side QD.

328·3
100·0
428·3 = QF + FD.
228·3 = QF — FD.

180
99
2) 81
40° 30′ = half the sum of the angles at the base QD.

428·3 : 228·3 :: tan. 40° 30′ : the tangent of half the difference of the angles at the base QD.

log. tan. 40° 30′	9.931499
log. 228·3	2·358506
	12·290005
log. 428·3	2·631748
log. tan. 19° 53′..	9·558257

40° 30′
19 53

DQO = 60 23 = Angle ODE.

20 37 = Angle DQF.

sin. 60° ·· 23′ : 328·3 :: sin. 99° : QD = 373.

Log. sin. 99° = 9·994620.
log. 328·3 — 2·516271

12·510891
log. sin. 60° 23′ — 9·939195

log. 373.———— 2·571696

2)373.

186·5 = QE.

90° 0′
60 . 23

29 37 = angle EOQ.

sin. 29° 37′ : 186·5 :: sin. 90° : OQ = 377·38.

Log. sin. 90° = 10·000000
log. 186·5 —— 2.270679

12·270679
log. sin. 29° 37′ —— 9·693898

log. 377·38 ——— 2·576781

1070
377·38

692·62 = OB,

the radius sought. This process is slow, but it carries the reasoning along with it.

EXAMPLE.

Let the radii QC and OC be perpendicular to the chord AB = 2900 feet; the angle DAB = 128°; angle GBA = 114°.
Required the lengths of the radii AQ and OB.

Put $2c$ = JAQ = angle DAB — 90°.
$2s$ = JBO = angle GBA — 90°.

Then OCB = 45° + s, and CBJ = 45° — s = IBH, which will now be proved equal to HBG.

CBJ = 45 — s
JBO = $2s$
OBG = 90
} add =

135 + *s* = angle CBG, which taken from 180°, gives 45 — *s* = HBG.

Fig. 86.

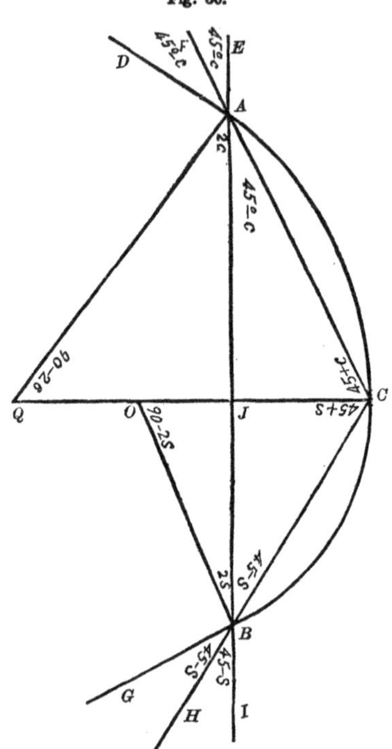

Hence, if the angle GBA be taken from 180 degrees, and the remainder divided by 2, the angle CBJ is found. In the same way it may be shown that the angle CAB = half EAD. A common tangent drawn at C would be parallel to the chord AB. And I have before shown that the angle ACB is always equal 90° + half QAB + half ABO.

$$\begin{array}{r} 180^\circ. \\ 128 \\ \hline 2)\ 52 \\ \hline \end{array}$$

angle CAB = 26° = angle FAE = FED.

180°
114
2) 66
33° = angle CBA.

128°
90
2) 38° = angle JAQ;
19.

114
90
2) 24° = JBO.
12.

90°
19
12
121° = angle ACB.

45°
19
64 = ACQ.

45°
12
57 = OCB.

sin. 121° : 2900 :: sin. 26° : CB = 1483·2

Log. sin. 26°	9·641842
log. 2900.........................	3·462398
	13·104240
Log. sin. 121° = log. sin. 59°.......	9·933066
log. 1483·2........................	3·171174

sin. 121° : 2900 :: sin. 33° : AC = 1842·6.

Log. sin. 33°.............	9·736109
log. 2900	3·462398
	13·198507
log. sin. 121°.............	9·933066
log. 1842·6	3·265441

90°
38
52 = AQC = DAE.

90°
24
66 = COB = IBG.

Sin. 52° : 1842·6 :: sin. 64° : QA = 2101·7

Log. sin. 64°	9·953660
log. 1842·6	3·265441
	13·219101
log. sin. 52°	9·896532
log. 2101·7	3·322569

Sin. 66° : 1483·2 :: sin. 57° : OB = 1361·5

Log. sin. 57°	9·923591
log. 1483·2	3·171174
	13·094765
Log. sin. 66°	9·960730
log. 1361·5	3·134035

Fig. 87.

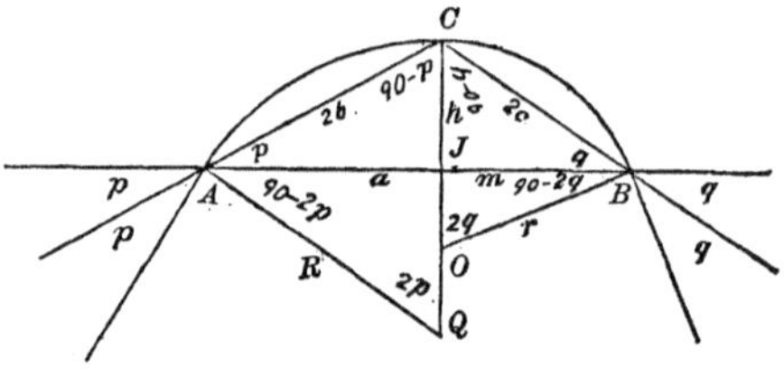

The radii required are thus found to be 1361·5 feet and 2101·7 feet, by the use of the table of logs. and log. sines only. But to generalize this example and involve other regular magnitudes, put AC = b; AB = $2a$; JB = m; AQ = R; OB = r; angle CAJ = p; CBJ = q, and, therefore, AQC = $2p$; COB = $2q$. Also put Bc = $2c$.

$$a - m = \text{AJ},$$

$$\frac{h}{a - m} = \tan. p. \therefore h = (a - m) \tan. p;$$

$$\frac{h}{m} = \tan. q \therefore h = m \tan q;$$

$$\therefore m \tan. q = a \tan. p - m \tan. p.$$

$$\therefore m = \frac{a \tan. p}{\tan. p + \tan. q}.$$

$$h = m \tan. q;$$

$$2b = (a - m) \sec. p;$$

$$2c = m \sec. q;$$

$$\frac{b}{\text{R}} = \sin. p \therefore \text{R} = \frac{b}{\sin. p}.$$

$$\therefore \text{R} = b \text{ cosec. } p.$$

$$r = c \text{ cosec. } q.$$

For the sake of illustration I will apply these formulas to the last numerical example.

$$a = 2900 \text{ feet}$$
$$p = 26^{\circ}$$
$$q = 33^{\circ}.$$

$$m = \frac{2900 \text{ tan. } 26}{\text{tan. } 26 + \text{tan. } 33} = \frac{\cdot 2900 + \cdot 4877}{\cdot 4877 + \cdot 6494} = 1243 \cdot 8$$

$$\begin{array}{r} 2900\cdot \\ 1243\cdot 8 \\ \hline 1656\cdot 2 \end{array} = a - m$$

$$\begin{aligned} \text{Log. } m &= 3\cdot 094751 \\ \text{Log. sec. } q &= \cdot 076409 \\ \hline & 3\cdot 171160 = \text{log. } 1483. \end{aligned}$$

$$2c = 1483 \text{ nearly.}$$

$$\begin{aligned} \text{Log. } (a - m) &= 3\cdot 219113 \\ \text{log. sec. } p &= \cdot 046340 \\ \hline & 3\cdot 265453 = \text{log. } 1842\cdot 6 \end{aligned}$$

$$26 = 1842\cdot 6 \text{ feet.}$$

$$\begin{aligned} \text{Log. } b &= 2\cdot 964423 \\ \text{Log. cosec. } p &= \cdot 358158 \\ \hline & 3\cdot 322581 = \text{log. } 2101\cdot 7 \end{aligned}$$

$$\therefore R = 2101.7 \text{ feet.}$$

$$\begin{aligned} \text{Log. } c &= 2\cdot 870130 \\ \text{Log. cosec. } q &= \cdot 263891 \\ \hline & 3\cdot 134021 = \text{log. } 1361\cdot 5 \end{aligned}$$

$$\therefore r = 1361\cdot 5 \text{ feet.}$$

PROBLEM.

Given the distance AB = a, Fig. 38, the tangent BD = b, and the tangent AD = c, to find the tangent AR = RC, which call x, and the tangent BS = SC, which call z; supposing the two branches of the compound curve to have the common tangent RS parallel to AB.

The triangles DRS and DAB are similar.

$$\therefore DR : RS :: DA : AB, \text{ that is,}$$

$$c - x : x + z :: c : a, \quad \text{(I.)}$$

$$\therefore x + z = \frac{ac - ax}{c} = a - \frac{ax}{c}$$

Fig 38.

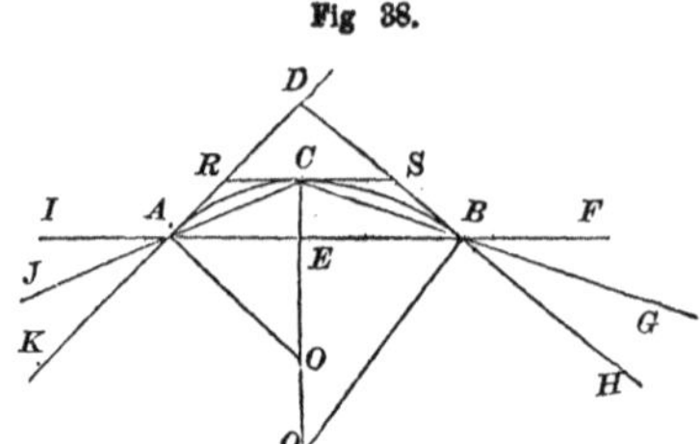

DS : SR :: DB : BA, that is,

$$b - z : x + z :: b : a, \quad \text{(II.)}$$

$$\therefore x + z = \frac{a(b - z)}{b} = a - \frac{az}{b}.$$

Hence,

$$a - \frac{ax}{c} = a - \frac{az}{b},$$

$$\therefore bx = cz. \quad \text{(III.)}$$

From the above proportions it appears that,

$$cx + cz = ac - ax, \text{ (I.}$$

$$\therefore cx + bx = ac - ax$$

$$\therefore x = \frac{ac}{a + b + c}$$

From (II.) $bx + bz = ab - az$, from (III.) substitute the value $bx = cz$.

$$\therefore cz + bz = ab - az$$

$$z = \frac{ab}{a + b + c}$$

EXAMPLE.

Let the distance AB = 2000 feet, Fig. 38, the tangent BD = 1600 feet, and the tangent AD = 1400 feet, what are the lengths of the parts RC and CS of common tangent when it is parallel to AB?

$$\text{RC} = \frac{2000 \times 1400}{2000 + 1600 + 1400} = 560 \text{ feet}$$

$$\text{CS} = \frac{2000 \times 1600}{2000 + 1600 + 1400} = 640 \text{ feet.}$$

The whole tangent RS = 1200 feet.

PROPOSITION XII.

In a Serpentine Curve, one Radius and its Tangential Point being given, to find the other Radius and the Closing Tangential Point.

From the given tangential point C, draw the given radius CO perpendicular to the tangent DR. Draw the curve CG to some point G, where it is found convenient that it should have its point of contrary flexure. Through OG draw the normal OGQ; from G draw GT perpendicular to OGQ to meet the tangent AT; make TB = TG, and draw BQ perpendicular to AT, meeting OGQ in Q; then Q is the required centre, and QB = QG is the radius of the curve BG, this property needs no demonstration, it is self evident from the nature of the circle and its tangents.

PROBLEM.

When the Tangential Points B and C, and the Radius OC, Fig. 39, are given, to find the other Radius QB.

Fig. 39.

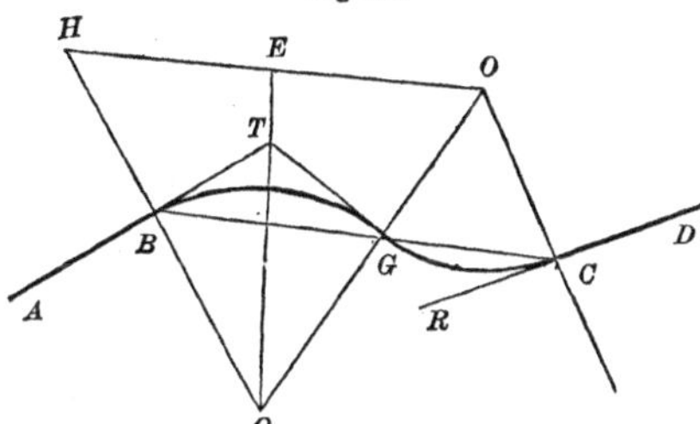

From the tangential points B and C, draw CO and BH respectively perpendicular to the tangents CD and AB, and equal to the given radius; join OH, and bisect it in F; draw FQ perpendicular to OH, meeting HB prolonged in Q; join Q and O; make QG = QB; then Q is the centre, and BQ = QG is the radius of the portion BG of the curve. BQ is the radius required.

EXAMPLE.

Let BC = a = 200 chains; the given radius OC = 110 chains = r; the required radius QB = x; m = 45° = angle TBC;
n = 20° = angle BCR.

(Fig. 39.) It will be presently shown that

$$x = \frac{a\,(a - 2\,r\,\text{sin.}\,n)}{2\left(a\,\text{sin.}\,m + 2\,r\,\text{sin.}^2\,\frac{m-n}{2}\right)}$$

$$\text{sin.}\,m = \cdot70711;\ \text{sin.}\,n = \cdot34202;$$

$$\text{sin.}\,\frac{m-n}{2} = \text{sin.}\,12^\circ\,30' = \cdot21644;$$

$$\therefore\ QB = x = \frac{200\,(200 - 220 \times \cdot34202)}{2\,(200 \times \cdot70711 + 220 \times \cdot21644^2)} =$$

$$\frac{200 \times 124\cdot7556}{2 \times 151\cdot7282} = 82\cdot22 \text{ chains.}$$

PROBLEM.

The tangential points B and C, Fig. 40, and their distance BC, being given; supposing the radii of the two portions of the serpentine curve to be of the same length, it is required to find them.

Fig. 40.

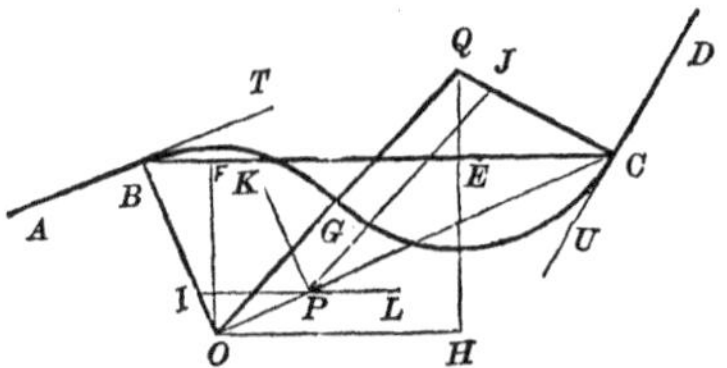

Draw BO perpendicular to AT, and QC perpendicular to UD; take JC = BI of any convenient length; draw IL parallel to BC; from J lay off JP = 2CJ = 2IB. The point P being thus found, draw CP and produce it to meet BO in O; and through O parallel to PJ draw OGQ meeting CQ in Q; then O and Q are the centres, and OB and QC are the radii of the serpentine curve BGC, the common normal of the portions BG, GC of the curve being OGQ = twice OB = twice CQ.

To prove this, draw PK parallel to OB; then by similar triangles,

CJ : JP = 2JC :: CQ : OQ = 2CQ;
∴ OQ = twice CQ.
IB = KP = JC;
CP : CO :: CJ = KP : QC,
CP : CO :: KP : OB,
∴ QC = OB, hence, OQ = twice OB.

EXAMPLE.

Let BC $= a =$ 200 chains; $m =$ angle TBC $= 27°$; $n =$ angle BCU $= 50°$; what is the length of the radius, $r =$ OB $=$ OG $=$ GQ $=$ QC.

It will be presently shown that

$$r = \frac{a}{\text{sin. } m + \text{sin. } n + 2 \text{ sin. of arc to cos. } \frac{1}{2} (\text{cos. } m + \text{cos. } n)}$$

From a table of natural sines, it is found that

$$\begin{array}{ll} \text{sin. } m = \cdot 45399, & \text{cos. } m = \cdot 89101; \\ \text{sin. } n = \cdot 76604, & \text{cos. } n = \cdot 64279; \\ \quad 1 \cdot 22003 & \quad 2)1 \cdot 53388 \\ & \quad \cdot 76690 \end{array}$$

$$\therefore \tfrac{1}{2} (\text{cos. } m + \text{cos. } n) = \cdot 76690$$

and twice the sine of an arc to cosine ·76690 is 1·28334.

$$\therefore r = \frac{200}{1 \cdot 22003 + 1 \cdot 28334} =$$

$$\frac{200}{2 \cdot 50337} = 79 \cdot 89 \text{ chains.}$$

INVESTIGATION.

Draw QEH and OF perpendicular to AB; through O draw OH parallel to BC.

$$\begin{array}{ll} \text{OQ} & = 2r; \\ \text{BF} & = r \text{ sin. } m; \\ \text{OF} & = \text{HE} = r \text{ cos. } m; \\ \text{CE} & = r \text{ sin. } n; \\ \text{QE} & = r \text{ cos. } n; \\ \therefore \text{QH} & = r (\text{cos. } m + \text{cos. } n); \end{array}$$

$$\frac{\text{QH}}{\text{OQ}} = \tfrac{1}{2} (\text{cos. } m + \text{cos. } n) = \text{sin. QOH},$$

which becomes known, as the expression $\frac{1}{2}$ (cos. m + cos. n), is composed of known quantities; hence the other angle OQH, of the right angled triangle OQH, is also known, and may be thus expressed:

$$\text{sin. OQH} = \text{sin. of arc to cos. } \tfrac{1}{2} (\text{cos. } m + \text{cos. } n).$$

$$\begin{array}{l} \text{OH} = \text{FE} = 2\,r \times \text{sin. OQH}, \\ \quad = 2\,r \times \text{sin. of arc to cos. } \tfrac{1}{2} (\text{cos. } m + \text{cos. } n). \end{array}$$

$$\text{BC} = \text{BF} + \text{FE} + \text{EC} = al$$

$$\therefore a = r \text{ sin. } m + r \text{ sin. } n + 2\,r \text{ sin. of arc to cos. } (\text{cos. } m + \text{cos. } n)$$

$$\therefore r = \frac{a}{\text{sin. } m + \text{sin. } n + 2 \text{ sin. of arc to cos. } (\text{cos. } m + \text{cos. } n)},$$

the expression to be established.

When the radii are unequal, and one of them, $r =$ QC is given, and the other R $=$ BO is required. In this case OQ $=$ R $+ r$;

$$\text{BF} = \text{R} \sin. m;$$
$$\text{OF} = \text{HE} = \text{R} \cos. m;$$
$$\text{HQ} = \text{R} \cos. m + r \cos. n;$$
$$\text{FE} = \text{OH} = \sqrt{\text{OQ}^2 - \text{QH}^2};$$
$$\text{OH} = \sqrt{(\text{R} + r)^2 - (\text{R} \cos. m + r \cos. n)^2};$$
$$\therefore a = \text{R} \sin. m + r \sin. n + \sqrt{(\text{R} + r)^2 - (\text{R} \cos. m + r \cos. n)^2};$$

by transposing and squaring, and remembering that $\sin.^2 + \cos.^2 = 1$, the result will be

$$a^2 - 2a\,(\text{R} \sin. m + r \sin. n) = 2\text{R}\,r\,\big(1 - (\cos. m \cos. n - \sin. m \sin. n)\big) = 2\text{R}\,r\,\big(1 - \cos.\,(m - n)\big) = 4\text{R}\,r \sin.^2 \frac{m-n}{2},$$

$$\therefore \text{R} = \frac{a\,(a - 2r \sin. n)}{2\left(a + 2r \sin.^2 \frac{m-n}{2}\right)};$$ the formula before used, which was to be established.

PROBLEM.

To make a given deviation HQ from a straight portion of a line of Railroad AHD, by means of three curves BG, GQD, DC, in order to avoid an obstruction.

Fig. 41.

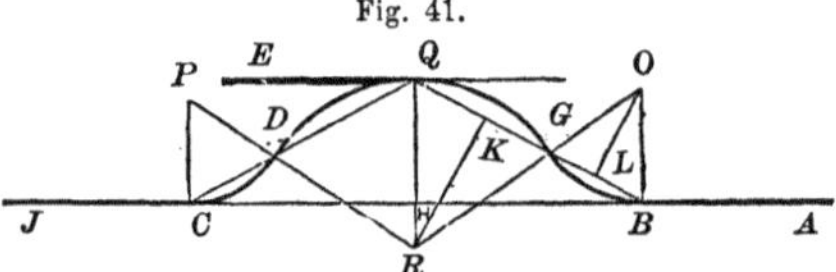

Take HQ any convenient distance from CB; make HB $=$ HC; CP, HQ, OB, are perpendicular to BC; with R as a convenient centre, describe the arc DQG, meeting QB and QC in G and D; draw RGO and RDP meeting the perpendiculars OB and CP in O and P.

Because QR $=$ RG, and OB parallel to QR, the angles RQG, RGQ, OGB, OBG are equal to one another; therefore OG $=$ OB, and consequently OB $=$ CP is the radius to connect the straight road AB and arc QG. It may be observed that the triangles RQG and OBG are similar, and that the reversing point G of a reversed curve between parallel tangents EQ and BA is always in the line QB joining the tangent points.

PROBLEM.

To make a given deviation HQ from a straight portion of a line of Railroad AHJ by means of three curves BG, GQD, DC, having their radii, PC, RQ, OB, all equal.

Perpendicular to JA, draw HQ = the given deviation. (Fig. 42.)

Fig. 42.

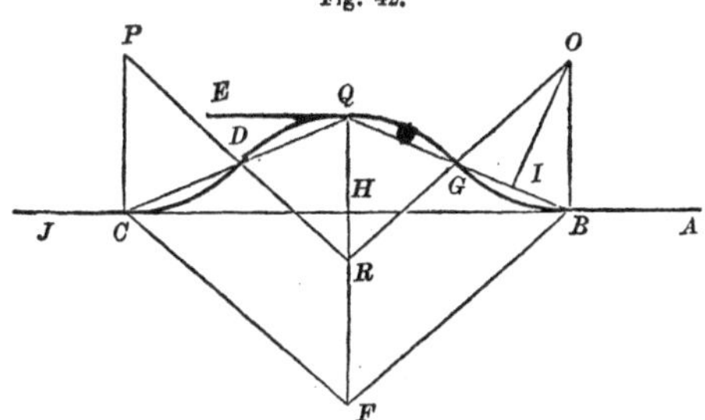

Make BF = FQ = twice QR; draw BO perpendicular to AB; through R, draw RGO parallel to FB, meeting OB in O; join B, Q, and C, Q, cutting RO and RP respectively in G and D. Then B and C are the starting and closing points of the curve; the chords BG, GQ, QD, DC are all equal, and so are the radii PC, RQ, OB.

On account of the parallels RF = OB = QR, for QR was made = RF; and RO = FB = QF = 2QR;

$$QF : FB :: QR : RG,$$

$$\text{but } QF = FB, \text{ therefore } QR = RG;$$

$$RO = 2QR = 2RG$$

$$\therefore RG = GO.$$

Consequently, the triangles OBG, RQC, RDQ, PCD are similar and equal. In Fig. 42, OI perpendicular to BG, then GI = IB; the right angled triangles OBI, QHB are similar;

$$\therefore OB : BI = \tfrac{1}{4}BQ :: BQ : QH,$$

$$\therefore \tfrac{1}{4}BQ^2 = (\tfrac{1}{2}BQ)^2 = GB^2 = BO \times QH.$$

$$\therefore GB = \sqrt{BO \times QH}.$$

$$HB^2 = BQ^2 - QH^2 = 4BO \times QH - QH^2 = 4QH\ (BO - QH)$$

$$\therefore HB = \sqrt{QH\ (4BO - QH)}.$$

EXAMPLE.

Let the given deviation HQ = 2 chains = n, and the common radius BO = 85 chains = R; required BH = HC; and GB = half BQ. Put a = BQ.

$$GB = \frac{a}{2} = \sqrt{R \times n} = \sqrt{170} = 13{\cdot}04 \text{ chains.}$$

$$HC = \sqrt{2\ (340 - 2)} = 26 \text{ chains.}$$

$$HC = \sqrt{n\ (4R - n)} = HB.$$

EXAMPLE.

Let the perpendicular distance HQ, Fig. 42, between two parallel tangents EQ, BA, be 20 feet $= n$, as in the last example, the distance between the two tangent points Q and B $=$ 60 feet, which put $= a$; required the common radius OB $=$ RQ, which put $=$ R.

$$\frac{a}{2} = \sqrt{R \times n},$$

$$\therefore R = \frac{a2}{4n} = \frac{60^2}{80} = 45$$

$$QG = GB = 30 \text{ feet.}$$

$$HB = \sqrt{n\ (4R - n)} = \sqrt{20\ (180 - 20)}$$

$$= \sqrt{3200} = 56{\cdot}568542 \text{ feet.}$$

EXAMPLE.

Given the perpendicular distance QH, Fig. 42, between two parallel tangents, EQ and BA $=$ 20 feet $= n$; required the distance BQ between the two tangents $= a$; the common radius R $=$ 45 feet.

$$a = 2\sqrt{R \times n} = 2\sqrt{45 \times 20} = 60$$

PROBLEM.

The line that joins the fixed tangent points being given $= a = {} = QB = 140$ feet; $n = 63$ feet $= QH$, the perpendicular distance between the two parallel tangents EQ, BA, Fig. 41; and the radius AE $= R = 88$ feet; being given to find $r = OB$; and the chords $QG = p$; and $GB = q$.

The triangles HQB and OLB are similar, and also QRK and QHB;

$$\therefore a : n :: R : \tfrac{1}{2}p;$$

$$\therefore \frac{2n\,R}{a} = p = QG.$$

$$\text{ang. } q = a - p = GB.$$

$$\frac{2 \times 63 \times 88}{140} = 79{\cdot}2 \text{ feet} = QG;$$

$$140 - 79{\cdot}2 = 60{\cdot}8 = GB.$$

The right-angled triangles QRK and GOL are also similar; hence,

$$a : n :: r : \tfrac{1}{2}q,$$

$$\therefore r = \frac{aq}{2n}$$

$$r = \frac{140 \times 60{\cdot}8}{2 \times 63} = 33{\cdot}77 \text{ feet.}$$

When R, r, n are given to find a, p, q. Taking the equations

$$p = \frac{2n\text{R}}{a};\ ap = 2n\text{R},$$

$$q = a - p;\ a = p \times q$$

$$r = \frac{aq}{2n};\ aq = 2nr.$$

By adding the first and third of these equations together,

$$aq + ap = a\,(p + q) = a^2 = 2n\,(\text{R} + r)$$

$$\therefore a = \sqrt{2n\,(\text{R} + r)};$$

$$p = \frac{2n\,\text{R}}{\sqrt{2n\,(\text{R} + r)}};\qquad q = \frac{2n\,r}{\sqrt{2n\,(\text{R} + r)}}.$$

EXAMPLE.

Let R $= 60$; $r = 40$; and $n = 18$;
Required a, p, q.

$$a = \sqrt{36\,(60 + 40)} = 60.$$

$$p = \frac{18 \times 60}{60} = 18,\quad q = \frac{18 \times 40}{60} = 12.$$

Any three of the six quantities R, r, n, a, p, q, being given, the other three may be found.

PROBLEM.

Given the length of the common tangent $TC = 800$ feet $= t$; the angle of intersection $ETG = 82^\circ = 2\theta$, angle of intersection $BCD = 62^\circ = 2\phi$; to find the common radius $AQ = QG = OG = OB = r$; to unite the tangents AT, BC, in an inversed manner.

$$\theta = \text{TQG} = 41^\circ;$$

$$\phi = \text{GOC} = 31^\circ;$$

$$\text{TG} = r \tan.\ \theta;\ \text{CG} = r \tan.\ \phi$$

$$\therefore \text{TC} = t = r \tan.\theta + r \tan.\phi;$$

$$\therefore r = \frac{t}{\tan.\theta + \tan.\phi} = \frac{800}{\cdot 60086 + \cdot 86929}$$

$$= \frac{800}{1\cdot 47015} = 544\cdot 16 \text{ feet.}$$

Fig. 43.

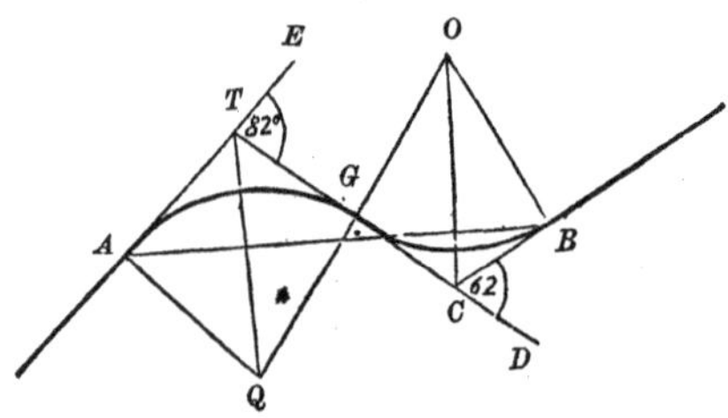

PROBLEM.

The distance between the fixed tangent points A, B, = a = 1270 *feet; the angle ABE =* 90° — θ = 76°; *the angle BAF =* 90° — ϕ = 30°; *find the common radius OB = r, to unite the tangents BE, AF, by a reversed curve.*

$$\theta = 90 - 76 = 14° = \text{OBD};$$
$$\phi = 90 - 30 = 60° = \text{QAC}.$$

Fig. 44.

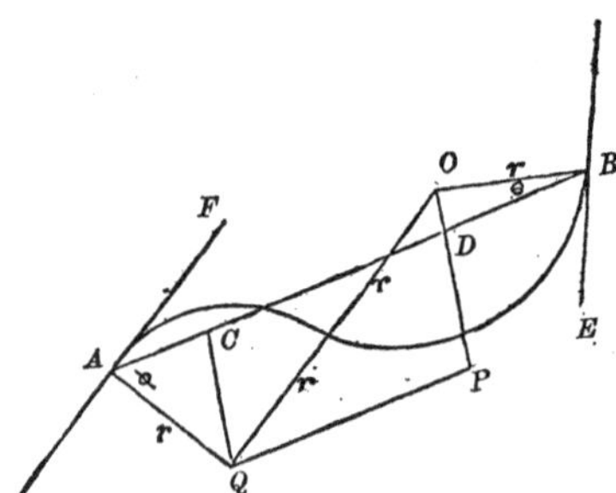

Draw QC and OD perpendicular to AB; and QP parallel to it.

$$OQ = 2r.$$

Nat. sin. 60° = ·8660254
nat. sin. 14° = ·2419219

2)1·1079473

·5539736 = natural sine

of 33° 38′⅖ = angle OQP.

Cosine 33° 38′⅖ = ·8325
" 60° = ·5000
" 14° = ·9703

These numbers will be required presently.

$$OP = r \sin.\ \theta + r \sin.\ \phi\ ;\ \frac{OP}{OQ} = \sin.\ OQP$$

$$= \frac{\sin.\ \theta + \sin.\ \phi}{2},$$ which is composed

of known quantities, and consequently the angle OQP is readily found, which call n, then.

$$2r \cos.\ n = PQ = DC.$$

$$AC = r \cos.\ \phi\ ;\ DB = r \cos.\ \theta$$

$$AB = a = r \cos.\ \phi + r \cos.\ \theta + 2r \cos.\ n$$

$$\therefore r = \frac{a}{\cos.\ \theta + \cos.\ \phi + 2 \cos.\ n}.$$

$$r = \frac{1270}{\cdot9703 + \cdot5000 + 1\cdot6650} =$$

$$\frac{1270}{3\cdot1353} = 405 \text{ feet},$$

the common radius required.

MISCELLANEOUS PROBLEMS

IN LAYING OUT RAILROAD CURVES, TURNOUTS, CROSSINGS, FROG ANGLES, &c.

PROBLEM.

Suppose AL and GE to be two Parallel Tangents to be united by a reversed curve of given common radius = 200 *feet* = r. *This curve has to pass through the point C, near A, and between the parallels; let CE be the direction of the tangent at the point C to make a given angle CEG* = 12° = 2θ, *with the track GE.*

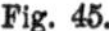

Fig. 45.

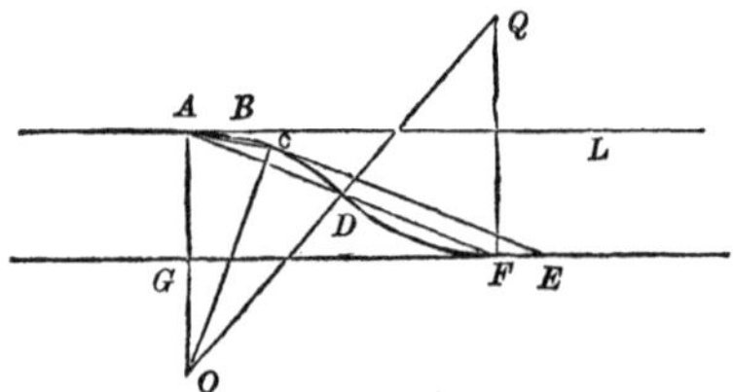

Since angle LBE = CEG = 2θ; BAC = BCA = θ, on account of the equal tangents BA, BC; hence the point A is readily found as

$$AC = 2T \sin. \theta.$$

Lay of the angle BCA = θ, and measure on AC the length $2r$ sin. θ; then measure the distance AG between tho parallels, which put = n = 20 feet in this particular case.

It has been before shown that

$$AF = a = 2 \sqrt{r \times n} = \sqrt{200 \times 20}$$

63·246 feet; and because AD = DF, AD or DF = 31·623 feet.

The direction of AF is also easily found, for

$$\frac{AG}{AF} = \frac{20\cdot}{63{\cdot}246} = \text{natural sine of the angle AFG} = \text{LAF}.$$

$$\frac{20\cdot}{63{\cdot}246} = {\cdot}31622 = \sin.\ 18^\circ\ 26'.$$

This problem is often used in placing switches, frogs, turnouts, and in connecting parallel lines of rails.

PROBLEM.

Let AF and EB be two parallel tracks, B and D fixed tangental points; BD $= a =$ 400 feet; the angle IDB $= 20° = \theta$; the

Fig. 46.

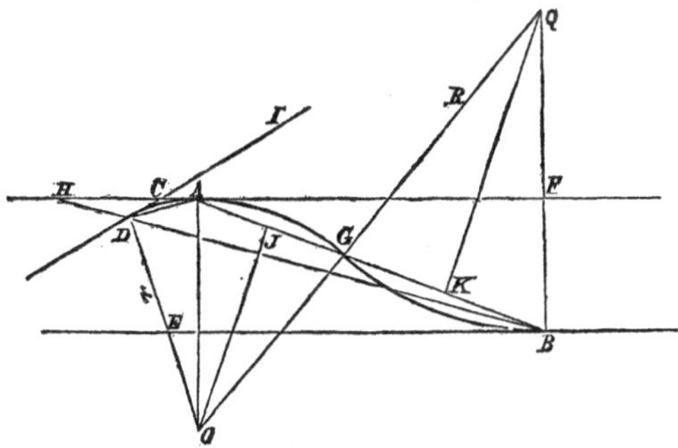

angle DBE $= 16° = \phi$; the radius OD $= r =$ 120 feet; required the radius QB $=$ R of the reversed curve DAGB.

On account of the parallels,

$$\text{angle CHB} = \text{HBE};\ \text{IDB} = \text{CHD} + \text{HCD};$$
$$\because \theta = \phi + \text{HCD}.$$

Because, on account of the equal tangents, angle CAD $=$ CDA,

$$\therefore \text{HCD} = \text{ICA} = 2\ \text{CDA}\ \theta - \phi$$

$$\therefore \text{CDA} = \frac{\theta - \phi}{2} = \frac{20 - 16}{2} = 2° \text{ becomes known.}$$

$$\text{AD} = 2\text{R sin.} \frac{\theta - \phi}{2}$$

AD $= 240 \times$ sin. $2° = 240 \times \cdot0349 = 8\cdot376$ feet. Then the point A is easily found by laying off the angle IDA $= 2°$ and DA $= 8\cdot376$ feet. Let AE be measured, which in this problem suppose $=$ 25 feet, $= n$. The triangles ABE, KQB, OJA are similar.

$$\therefore \text{AB} : \text{AE} :: \text{OA} : \text{AJ}, \text{ that is}$$
$$400 : 25 :: 120 : 7\tfrac{1}{2} = \text{AJ}.$$
$$\therefore \text{AG} = 15, \text{ and GB} = 400 - 15 = 385.$$

$$\text{GK} = \frac{385}{2} = 192\tfrac{1}{2}$$

$$\text{AE} : \text{AB} :: \text{KG} : \text{GQ} = \text{R, that is,}$$

$$25 : 400 :: 192\tfrac{1}{2} : 3080 \text{ feet the radius required.}$$

PROBLEM.

Let AB represent a Switched Rail from the main track AD, let CB = 6 inches = q' and AB = 24 feet = p (Fig. 47); the gauge of the road CG = 4·7 feet = g; the Radius OB = 100 feet = r. Required the Chord BF = c, and the Frog-angle HFI = f.

Fig. 47.

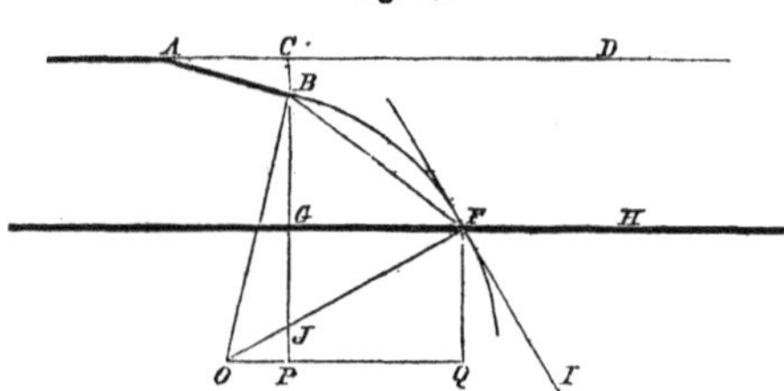

Draw OQ parallel to GH, and CP and FQ perpendicular to it. AB and FI being tangents at the points B and F, the angles HFI and QFO are equal, and the triangles ABC, BOP are similar.

$\frac{\cdot 5}{24} = \cdot 0208333$ = natural sine of the angle CAB = angle OBP $= 1^\circ\ 11 \cdot 625 = n$; cosine $= \cdot 9999943$.

$$\cos. f = \frac{FQ}{OF} = \frac{BP - BG}{r} = \frac{r \cos. OBP - (g - q)}{r}$$

$$= \frac{100 \cdot \cos. OBP - 4 \cdot 2}{100} = \frac{99 \cdot 99943 - 42}{100} = \cdot 9579943$$

$$\therefore f = HFI = 16^\circ\ 40' \text{ nearly.}$$

In general terms,

$$AC = \sqrt{p^2 - q^2},$$

$$p : \sqrt{p^2 - q^2} :: r : \frac{r\sqrt{p^2 - q^2}}{p} = BP.$$

$$\frac{r\sqrt{p^2 - q^2}}{p} - (g - q) = FQ = GP.$$

$$\cos. f = \frac{\sqrt{p^2 - q^2}}{p} = \frac{g - q}{r}$$

To find the chord BF = c.

$\frac{BG}{BF}$ = sin. BFG = sin. $\frac{1}{2}(n + f)$ as will be presently shown.

$$\therefore \frac{g - q}{c} = \sin. \tfrac{1}{2}(n + f)$$

$$\text{and } c = \frac{g\,q}{\sin. \tfrac{1}{2}(n + f)}.$$

$$\therefore c = \frac{4{\cdot}7 - {\cdot}5}{\sin. (8^{\circ}\ 56')} = \frac{4{\cdot}2}{{\cdot}15529}$$

$= 27{\cdot}05$ feet $=$ BF the required chord.

By the nature of tangents and on account of the parallels,

$$\text{Angle GJF} = \text{JFQ} = \text{HFI} = f$$
$$\text{GJF} = \text{OBJ} + \text{JOB}$$
$$\text{JOB} = f - n;$$
$$\text{Diff. of} \left\{ \begin{array}{ll} 90 - \tfrac{1}{2}(f - n) = & \text{BFO} \\ 90 - f \quad\quad\quad = & \text{GFO} \end{array} \right\}$$
$$\tfrac{1}{2}(f + n) = \text{angle BFG.}$$

$\dfrac{\sqrt{p^2 - q^2}}{p}$ = sine of ABC or cosine of the known angle CAB, which I put $= n$.

$$\therefore \cos. f = \cos. n - \frac{g - q}{r}.$$

From this equation, the frog angle f being given, the radius r may be found, for

$$r = \frac{g - q}{\cos. n - \cos. f}$$

and it has been shown that,

$$c = \frac{g - q}{\sin. \tfrac{1}{2}(n + f)}$$

EXAMPLE.

Let the frog-angle $f = 36^{\circ}$; the gauge of the road, $g =$ 4ft. 8½ in.; the length of the rail switched, $p =$ 18 feet; the range of the switched rail $q =$ 8½ inches; required the radius OB $= r$, and the chord BF $= c$, Fig 47.

18 feet = 384 half inches.

8½ in. = 17 half inches.

$$\frac{17}{384} = {\cdot}0442708 = \sin. 2^{\circ}\ 32'{\cdot}242$$

$$\begin{array}{lr} \text{cosine } 2^{\circ}\ 32'{\cdot}242 = & {\cdot}9990196 \\ \cos. 36^{\circ} \quad\quad\quad = & {\cdot}8090170 \\ \hline & {\cdot}1900026 \end{array}$$

$$\begin{array}{ll} g = & \text{4ft. } 8\tfrac{1}{2}\text{in.} \\ q = & \quad\ \ 8\tfrac{1}{2} \\ \hline & 4 \quad = g - q \end{array}$$

$\dfrac{4}{{\cdot}19}$ 21.05 feet $= r$

$$\begin{array}{rr} 36^\circ & 0' \\ 2 & 32{\cdot}242 \\ \hline 2)\ 38 & 32{\cdot}242 \\ \hline \end{array}$$

$$\sin.\ 19\quad 16{\cdot}121 = {\cdot}3299986.$$

$$c = \frac{g - q}{\sin.\ \frac{1}{2}\,(n + f)} = \frac{4\cdot}{{\cdot}3299986} = 12{\cdot}12 \text{ feet.}$$

PROBLEM.

Let AB, CD, and GH, be three straight parallel tracks; AF a turnout as far as the frog F; the Frog-angle DFI = and 2θ = 44°, and the distance DH = m = 32 feet, being given; to find the second Radius OF the reversing point being at F.

Fig. 48.

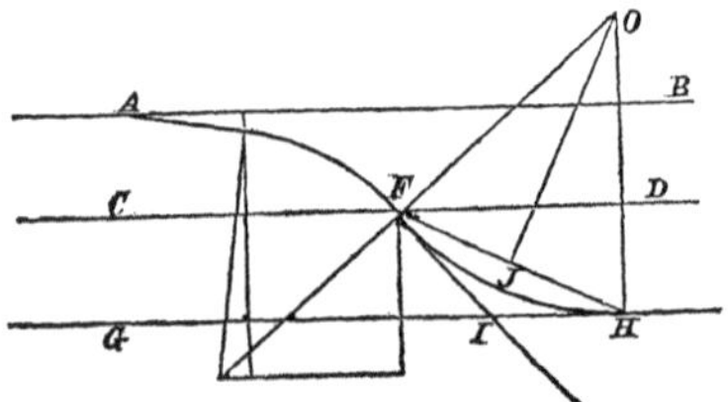

The frog-angle 2θ = DFI = FIG = 2FHI, because IF = IH.

∴ OHF = $90^\circ - \theta$ = OFH; also, angle DFH = FHI = θ.

$$\sin.\ \theta : \text{DH} :: \sin.\ 90^\circ : \text{FH, that is,}$$

$$\sin.\ \theta : m :: 1 : \frac{m}{\sin.\ \theta} = \text{FH};$$

$$\text{JOH} = \text{FHI} = \theta$$

$$\text{Sin.}\ \theta : \tfrac{1}{2}\text{FH} :: \sin.\ 90^\circ : \text{OH}$$

$$\sin.\ \theta : \frac{\frac{1}{2}m}{\sin.\ \theta} :: 1 : \frac{\frac{1}{2}m}{\sin.^2\ \theta} = \text{OH, the radius required.}$$

Log. $\frac{1}{2}$ m	1·204120
log. sin. $^2\theta$ = 2 log. sin. θ............	19·147150
log. 114·02	2·056970 rejecting

20 in the index for the true log. sine of 22° = $\bar{1}{\cdot}573575$, the tabular log. is set down 10 more than this to avoid the use of negative indices.

OH = 114 feet nearly.

Log. m = log. 32	1·505150
log. sin. θ	9·573575
log. 85·423	1·931575

FH = 85·423 feet.

PROBLEM.

Let KMGBE be a Crossing from the straight track KJ to a parallel one CE, Fig. 49. Let KM and BE represent the switched rails of equal length. Measure LC perpendicular to KMH, so that the angles LMK and DBE are right angles, measure also the distance CD = AB.

Fig. 49.

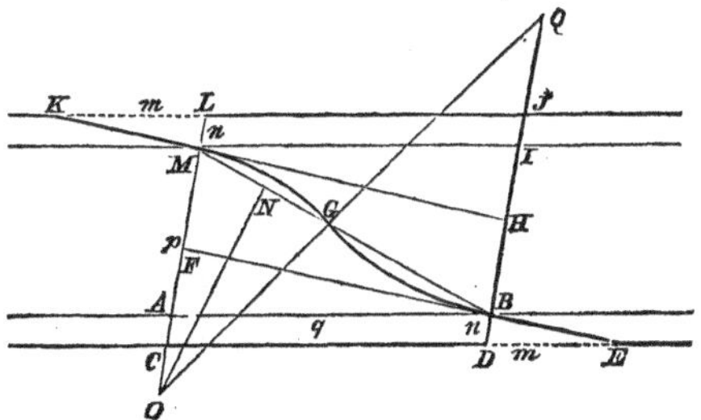

Given OM = R,
KL = DE = m,
LM = BD = n,
DC = AB = q,
MA = IB = p.

To find GQ = r; MG = x, GB = y:

Because LKM = BED, the lines KH and EF are parallel. MH is perpendicular to AM, and BF to JB, hence BHMF is a rectangular parallelogram.

$$m : n :: q : \frac{nq}{m} = \text{AF}$$

$$p - \frac{nq}{m} = \text{MF} = \text{BH}.$$

It has been shown in one of the foregoing problems that the triangles MNO and MHB are similar,

$$\therefore \text{R} : \frac{x}{2} :: x + y : \text{BH} =$$

$$\therefore x = \frac{2\,R \times BH}{x + y} = \frac{2\,R \times \left(p - \frac{nq}{m}\right)}{x + y}$$

$$x + y = MB,$$

$$q^2 - \frac{n^2q^2}{m^2} = FB^2, \text{ and}$$

$$MF^2 + FB^2 = MB^2 = (x + y)^2, \text{ that is,}$$

$$\left(\frac{p - nq}{m}\right)^2 + q^2 - \frac{n^2q^2}{m^2} = (x + y)^2$$

$$\therefore x + y = \sqrt{p^2 - \frac{2npq}{m} + q^2}, \text{ and}$$

$$x = \frac{2\,R \times \left(p - \frac{nq}{m}\right)}{\sqrt{p^2 - \frac{2npq}{m} + q^2}}.$$

Consequently, y becomes known also, and

$$r = \frac{xy}{p - \frac{nq}{m}}.$$

Let $m = 200$ inches;
$n = 10$ inches;
$q = 200$ feet $= 2400$ inches;
$p = 160$ inches.
$R = 150$ feet $= 1800$ inches

$$p - \frac{nq}{m} = 160 - \frac{10 \times 2400}{200}$$

$$= 160 - 120 = 40 \text{ inches}$$

$$x = \frac{3600 \times 40}{\sqrt{160^2 - 38400 + 2400^2}} = \frac{144000}{2397 \cdot 3}$$

$$60 \cdot 7. \quad 2397 \cdot 3 - 60.\ 7 = 23376 = y.$$

$$r = \frac{2337 \cdot 6 \times 60 \cdot 7}{40}.$$

PROBLEM.

When both Radii are equal, it is required to find them, m, n, p, q, being given as in the last problem.

In this case, $x = y$, and

$$r = \frac{(x+y)^2}{4 \times \text{BH}} = \frac{p^2 + q^2 - \dfrac{2npq}{m}}{4\left(p - \dfrac{nq}{m}\right)}.$$

$$x + y = 2x = 2y = 2\sqrt{r \times \left(p - \frac{nq}{m}\right)}.$$

By introducing angular magnitudes, this problem may be solved in the succeeding manner.

Fig. 50.

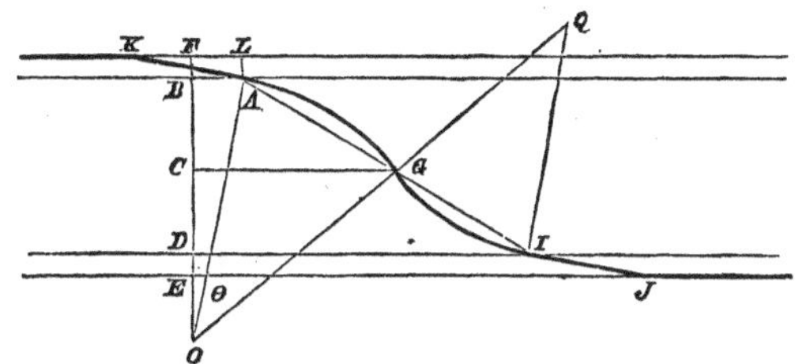

The length of the switch rail AK = IJ = say 20 feet; BF = LA = DE = 10 inches, hence the angle AKL = BOA, may be readily found, for

$$\frac{\text{AL}}{\text{KA}} = \frac{10}{20 \times 12} = {\cdot}0416667 =$$

natural sine of 2° 23′·2812. In making these calculations, I employ my "Calculators' Constant Companion," published by J. W. Moore, of Philadelphia.

OF is perpendicular to EJ, DJ, AB, KL, and CG is parallel to these lines.

Let BD = EF = twice FB = 30 feet, which put $= n$. Put θ = angle AOG, and c = AOB = 2° 23′·2812, found above.

$$\text{Then, OC} = \text{R cos.}\ (\theta + c) = \text{OB} - \text{BC}.$$
$$\text{OB} = \text{R cos.}\ c.$$

Now BC has to be found in known terms,

$$\text{BC} : \text{CD} :: \text{AG} : \text{GI} :: \text{R} : r;$$
$$\therefore\ \text{BC} : \text{BC} + \text{CD} :: \text{R} : \text{R} + r;$$
$$\text{but BC} + \text{CD} = \text{DB} = 30\text{ft.} = n;$$
$$\therefore\ \text{BC} = \frac{n\,\text{R}}{\text{R} + r},$$
$$\therefore\ \text{R cos.}\ (\theta + c) = \text{R cos.}\ c - \frac{n\,\text{R}}{\text{R} + r}.$$

and hence,

$$\cos.\ (\theta + c) = \cos.\ c - \frac{n}{R + r};$$

and $\theta + c$ becomes known and consequently θ. Let R = 500 feet, and $r = 300$, as they are supposed to be given; when the radii are equal the last equation will stand thus:

$$\cos.\ (\theta + c) = \cos.\ c - \frac{n}{2R}.$$

Natural cosine c = ...	·9991316
$\frac{n}{R + r} = \frac{30}{800}$ = ...	·0375000
cosine 15° 55′·3754	·9616316

From $\theta + c$ = 15° 55′·3754
Take c = 2 23·2812

θ = 13 32·0942

The triangles OAG and QGJ being similar, the angle GQJ = θ as well as AOG.

$$AG = 2\,R \sin.\frac{\theta}{2} \text{ and } IG = 2\,r \sin\frac{\theta}{2}.$$

$$\sin.\frac{\theta}{2} = \sin.\ (6^\circ\ 16'{\cdot}0471) = {\cdot}1091696.$$

·1091696	·1091696
1000	600
AG = 109·1696000;	IG = 65·5017600

Suppose R = r = 500, then
AG = GI, may be found as follows:

cosine c	·9991316
$\frac{n}{2R} = \frac{30}{1000}$	·0300000
cos. $(\theta + c)$ =	·9691316

$\theta + c$ = 14° 16′·3863
c = 2 23·2812

θ = 11 53·1051

$\frac{\theta}{2}$ = 5° 56′·55255

sin. 5° 56′·55255 = ·0861622
1000

$$AG = 2R \sin.\frac{0}{2} = 86{\cdot}1622000$$

$\therefore$ AG = GI = 86·1622 feet.

CUTTINGS AND EMBANKMENTS.

SIDE DEPTHS AND SIDE STAKES.

When the centre stumps of a railroad have been put down, which are usually at the distance of one chain, the line must next be levelled, and the number of the stumps entered in the level-book, in a vertical column; and opposite each number, in another column, the depth of the cuttings or embankments; and in a third column, the horizontal half-width of the surface cuttings. But every engineer has his peculiar method of keeping a field or level-book.

PROBLEM.

To set out the Width of Cuttings, when the surface of the ground is laterally level, and at a given height above the level of the intended Railroad, the ratio of e Slopes being given.

Fig. 51.

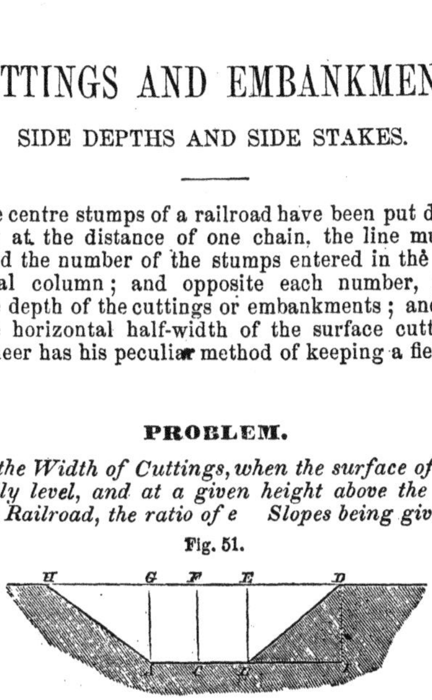

Let ABDH, Fig. 51, be the cross section of a cutting, the ground HD parallel with the bed of the road AB; put CF, the height in the centre of the road $= h$, and the breadth of the road-bed AB $= b$; the slope of the side AH or BD is generally expressed by the ratio of the base BI to the perpendicular ID; let BI to ID $=$ m to n, then the slopes are said to be m to n or $\frac{m}{n}$.

Then, as it is supposed, the ground HD is level, the distance from the centre F, to the side stakes D and H will be expressed by

$$\tfrac{1}{2} b + \frac{m}{n} h.$$

EXAMPLE.

Let the bottom width AB $=$ 28 feet, the depth of the cutting CF $=$ 16 feet, the slopes as 5 : 4, that is, BI : ID $=$ 5 : 4; required the distance of the side stakes H and D from the centre F.

$$HF = \tfrac{1}{2} \times 28 + \tfrac{5}{4} \times 16 = 34 = FD.$$

EXAMPLE OF EMBANKMENT, WHEN THE SURFACE OF THE GROUND IS LATERALLY LEVEL. (FIG. 52.)

Suppose the breadth of the roadway AB = 32 feet; the height of the embankment CF = 20 feet, and the ratio of the slopes 6 to 5, that is DE to EB = 6 : 5, required the distance of the side stakes H and D from the centre F.

Fig. 52.

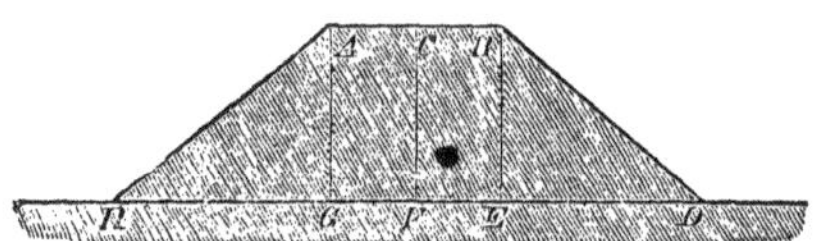

$$HF = \tfrac{1}{2} \times 32 + \tfrac{6}{5} \times 20 = 40 = FD.$$

The slope given to the sides of either cuttings or embankments varies with the material through which the road has to pass.

When BI = ID, the slope is said to be 1 : 1.

The slope is said to be a rise of 2 to 1, when BI = twice ID. In close jointed rock the ratio varies from 1 : 4 to 1 : 2.

In soft or loose-jointed rock, or stiff clay, the ratio varies from 1 : 1 to 3 : 2. If the road passes through moist springy ground, or loose sand, the ratio of the slope varies from 2 : 1 to 5 : 2.

Should the ground rise from F to M, Fig. 53 the slope stake must be set out further as at M; let the additional height

$$MN = p, \text{ then } DN = \frac{m}{n} p.$$

And if the ground falls from F to K, a distance LK = q, the slope stake must be set further in at K, a distance

$$HL = \frac{m}{n} q.$$

$$\therefore\ FN = \tfrac{1}{2} b + \frac{m}{n} h + \frac{m}{n} p$$

$$FL = \tfrac{1}{2} b + \frac{m}{n} h - \frac{m}{n} q$$

It often happens that p and q are equal.

Setting slope stakes for embankments resembles setting them for excavations, only a rise from the centre F, with excavations, corresponds to a fall with embankments; in fact, an embankment is a cutting inverted. Fig. 54 represents an embankment, and is Fig. 53 inverted.

Fig. 53.

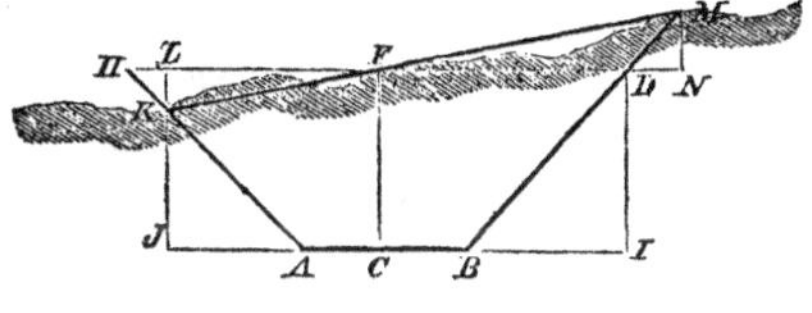

Fig. 54.

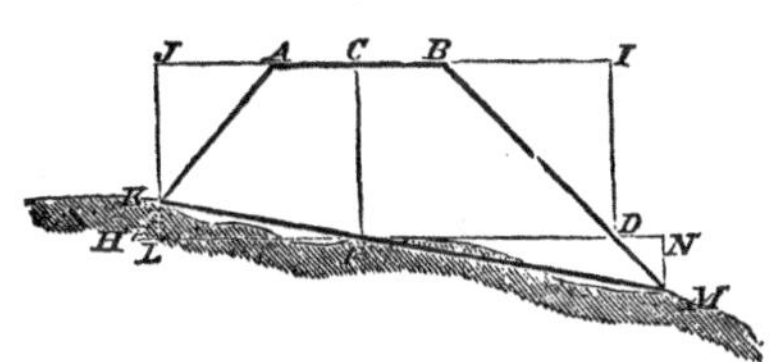

The rise at K is nearer the centre F, than H on the level, in an embankment, Fig. 54. While the fall at K is nearer the centre F than H on the level with F, in an excavation, Fig. 53.

EXAMPLE.

In the cutting, Fig. 53, and in the embankment, Fig. 54, let

$$\begin{aligned} AB &= b = 30 \text{ feet,} \\ CF &= h = 18 \text{ feet,} \\ MN &= p = 6 \text{ feet,} \\ KL &= q = 4 \text{ feet,} \end{aligned}$$

$$m \text{ to } n = 3 : 2, \text{ that is,}$$

$$3 : 2 = BI : ID = AJ : JK = DN : NM = HL : LK.$$

Consequently,

$$FN = \frac{30}{2} + \frac{3 \times 18}{2} + \frac{3 \times 6}{2} = 51 \text{ feet.}$$

$$FL = \frac{30}{2} + \frac{3 \times 18}{2} - \frac{3 \times 4}{2} = 30 \text{ feet.}$$

As the ground continues to slope up or down from F, the centre stake, the position of M and K are often determined on the ground by a series of trials, or fudged out in an office by some clumsy mechanical construction or other. To avoid guessing, or rule-of-thumb operations, I will lay down a practical exact plan by which the positions of the side stakes K and M may be easily found. The positions of D and H on a level with F, the centre stake, can be acurately calculated when the height CF, breadth AB, and ratio of BI to ID are given, therefore it is known, very nearly, where the lines AK and BM strike the surface of the earth. And as FK and FM are seldom in the same straight line, it is more accurate to find the ratio of FL to FK as well as the ratio of FN to NM. When these ratios are known, which may be readily found by a level and target-rod, the distance of M from F and of K from F are easily calculated. In the neighborhood of M and K, there are always short spaces before or behind M and K in the directions of the lines KF and FM, and it does not matter how irregular the surface is between these points. It is not necessary that the spaces before or behind K and M are level or not, so that they are nearly in the direction of FM and FK. Set up and adjust the level at any convenient place, X outside the cross section; place the target-staff at Y, as near as you are able to judge to the required point M; read off the height SY; remove the staff to the centre stake F, and read off the height FI; the difference between SY and FI will give QF, which put $= r$, measure YF, and put $s =$ YF. If the distance QY be not great, it may also be measured by the same tape or chain that takes the length of FY; or QY may be calculated for QY $= \sqrt{s^2 - r^2}$, which I will put $= t$ to make the reasoning more concise.

Fig. 55.

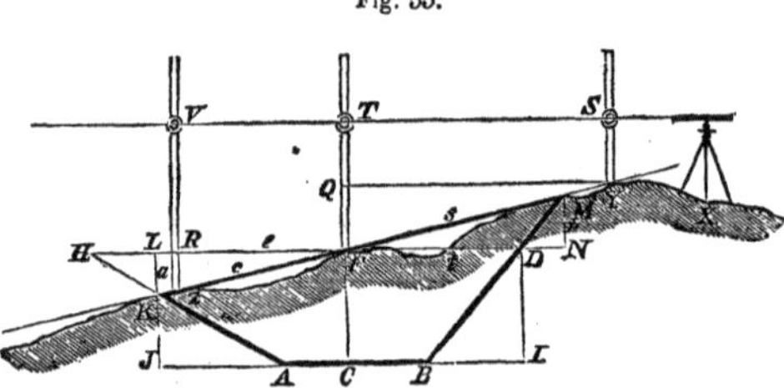

Then the three sides of the triangle QFY become known; this triangle is similar to the triangle FNM, and hence

$$\text{FN} : \text{NM} :: t : r; \text{ and}$$
$$\text{FN} : \text{FM} :: t : s; \text{ and}$$
$$\text{FN} : \text{MN} :: s : r.$$

Because the three sides of the triangle YF, QF, QY, are respectively represented by the three known quantities s, r, t.

Two of these quantities, s and r, may be measured in links or feet on the ground, and the third side may be measured or calculated according to the circumstance of the case. Again place the target-staff at Z, as near the required point K as you are able to judge; but it does not much matter where it is placed, as I have before observed, so that it is in the line FK.

Then read off the height ZV without changing the position of the instrument at X; from the height VZ take TF, the remainder

ZR is one of the sides of the triangle FRZ; put this known height ZR $= a$; measure ZF with a tape or chain, and put it $= c$; RF may be measured and put $= e$, or calculated, for $e = \sqrt{c^2 - a^2}$. As on the other side of the centre F, the triangle FRZ is similar to the triangle FLK. Hence,

$$\text{FK} : \text{KL} : \text{LF} = c : a : e.$$

To render this method of proceeding as clear as possible, I have dwelled on every point of the process, so that there could be no misunderstanding. This subject has been treated by other writers in a most slovenly manner, and by most of the empirical rules laid down by them, it takes three or four trials to determine the position of a side stake.

$$\text{Let KF} = x \text{ and FM} = y.$$
$$\text{AB} = b;\ \text{FC} = h.$$
$$\text{BI} : \text{ID} = m : n.$$

And we have just found by the level and target-staff that,

$$\text{FN} : \text{NM} = t : r;$$
$$\text{FM} : \text{MN} = s : t;$$

Also,

$$\text{FL} : \text{LK} = e : a;$$
$$\text{FK} : \text{KL} = c : a.$$

I have selected these measures in the most general manner in order that result may embrace all like cases.

$$c : a :: x : \frac{ax}{c} = \text{KL}.$$

$$\text{KL} : \text{LH} = \text{KJ} : \text{JA} = n : m,$$

$$\therefore \frac{ax}{c} : \text{LH} = n : m, \text{ or } \frac{amx}{cn} = \text{LH}.$$

$$\text{FH} - \text{HL} = \text{LF}.$$

$$\text{FH} = \tfrac{1}{2}b + \frac{m}{n}h;$$

$$\therefore \text{LF} = \tfrac{1}{2}b + \frac{m}{n}h - \frac{amx}{cn};$$

$$c : e :: x : \frac{ex}{c} = \text{LF}.$$

$$\therefore \frac{ex}{c} = \frac{b}{2} + \frac{mh}{n} - \frac{amx}{cn},$$

$$\therefore x = \frac{c\,(nb \times 2mh)}{2\,(en + am)}.$$

and hence the exact distance of the side stake K, from the centre F, becomes known.

$$s : r :: y : \frac{ry}{s} = \text{MN}.$$

$$\text{BI} : \text{ID} = \text{DN} : \text{NM} = m : n$$

$$\therefore\ n : m :: \frac{ry}{s} : \frac{mry}{ns} = \text{DN}.$$

$$\text{FD} + \text{DN} = \text{FN}.$$

$$\text{FD} = \tfrac{1}{2}\,b + \frac{m}{n}\,h;$$

$$\therefore \text{FN} = \tfrac{1}{2}\,b + \frac{m}{n}\,h + \frac{mry}{ns}.$$

$$s : t :: y : \frac{ty}{s} = \text{FN}$$

$$\therefore \frac{ty}{s} = \frac{b}{2} + \frac{mh}{n} + \frac{mry}{ns};$$

$$\text{and } y = \frac{s\,(nb + 2\,mh)}{2\,(tn - rm)}; \text{ and}$$

hence the exact distance of the side-stake M, is readily determined.

EXAMPLE.

Given the breadth of the roadway AB $= b = 28$ feet; the height of the centre stake CF $= h = 24$ feet; the ratio of the slopes AH, BM $=$ BI : ID $= m : n = 3 : 2$. In a surface distance, FY $= 60$ feet $= s$, a rise, FQ $= 7$ feet $= r$, is found by the level and target-staff; in the surface distance, FL $= 50$ feet $= c$; a fall, RZ $= 4$ feet $= a$ is found. Required the points K and M where the surface of the ground intersects the slopes that form the road.

The solution of this problem and this first example are given at full length, in order that the ground-work of the contracted practical rule may be better understood.

$$\text{FM} : \text{FN} : \text{MN} = s : \sqrt{s^2 - r^2} : r\cdot$$

$$\text{FK} : \text{FL} : \text{LK} = c : \sqrt{c^2 - a^2} : a.$$

$$60^2 = 3600$$
$$7^2 = 49$$

$$t = \sqrt{s^2 - r^2} = \sqrt{3551} = 59{\cdot}59.$$

$$50^2 = 2500$$
$$4^2 = 16$$

$$e = \sqrt{c^2 - a^2} = \sqrt{2484} = 49{\cdot}84.$$

$$x = \frac{50\ (2 \times 28 + 2 \times 3 \times 24)}{2\ (49{\cdot}84 \times 2 + 4 \times 3)} = \frac{10000}{223{\cdot}36} =$$

44·8 feet = the distance from F to K.

$$y = \frac{60\ (2 \times 28 + 2 \times 3 \times 24)}{2\ (59{\cdot}59 \times 2 - 7 \times 3)} = \frac{12000}{196{\cdot}36} =$$

61·11 feet = the distance from F to M. Although this calculation is simple and extremely accurate, yet the generality of practical men require rules that can be applied without entering into the reasoning of the matter in every particular case and example; to suit this class of practitioners I will lay down one or two other methods of finding the values of x and y.

Fig. 56.

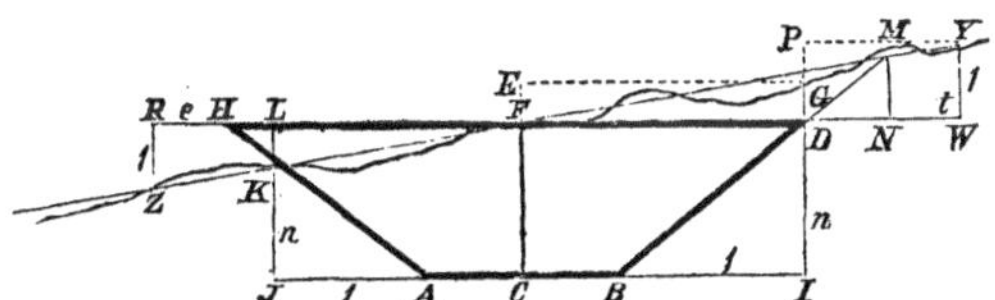

In Fig. 56, by the use of the level and target-staff, as described in Fig. 55, let the rise from the centre stake F, in the direction FY, be determined, which I will represent by the

ratio $t : 1$;

That is, FW : WY :: t : 1.

In the same manner, by placing the target-staff at Z, in the neighbourhood of K, find the fall from the centre stake F, so that FKZ may be in the same straight line, or nearly so; in general terms I will represent this ratio by $e : 1$, that is,

FR : Rz :: e : 1.

Before going on the ground to set out side-stakes, the lines of the figure HDBA are known; the horizontal half breadths FD = FH is found from the height FC, and breadth of roadway AB, being given.

When the positions of H and D are known, it is not difficult to select some points, Z and Y, near them to ascertain the slope of the ground.

In finding the half-breadths FD and FH, the ratio of BI to ID is also given; for my purpose at present I will represent this ratio by $1 : n$, that is,

BJ : JD = $1 : n$.

I need scarcely remark again, that

DN : NM = $1 : n$; HL : LK = $1 : n$.

Put v = HL, then,

$1 : n :: v : nv$ = LK.

Put HF $= d =$ FD, the half horizontal breadth, through the centre stake F.

$$e : 1 :: d - v : \frac{d - v}{e} = \text{KL also};$$

$$\therefore nv = \frac{d - v}{e}$$

$$\therefore v = \frac{d}{ne + 1}$$

Again put DN $= z$, then,

$$1 : n :: z : nz = \text{MN},$$

$$\text{and } t : 1 :: d + z : \frac{d \times z}{t} = \text{MN}$$

$$\therefore nz = \frac{d + z}{t}$$

$$\therefore z = \frac{d}{nt - 1}$$

From which the following simple practical rule may be deduced.

RULE.

When the ground rises from the centre, increase the horizontal half-breadth by the half-breadth divided by the product (less one) of the numbers that express the ratio of the rise and the ratio of the slope, and the horizontal distance of the side-stake is determined.

When the ground falls from the centre, decrease the horizontal half-breadth by the half-breadth divided by the product (plus one) of the numbers that express the ratio of the rise and the ratio of the slope, and the horizontal distance of the other side-stake is found.

EXAMPLE.

Given the breadth of the roadway AB $= 30$ feet; the height CF $= 26$ feet; the side slopes 1 : 2 (BI : ID :: 1 : 2); the rise of the surface from F to Y $= 1$ in 20 (FW : WY :: 20 : 1); the fall from F to Z $= 1$ in 36; required the horizontal distances FN and FL, where the surface of the ground intersects the side slopes of the railroad.

$$\text{H F} = \frac{30}{2} + \frac{1}{2} \times 26 = 28 = \text{FD, the half horizontal}$$

breadth meeting the side slopes on a level line through F.

The horizontal distance FW may be measured in lengths EG, PY, when the surface is very irregular.

$$2 \times 20 - 1 = 39.$$

```
39)28·0 (·72
   27 3
   ----
     70
     78
```

$$\therefore \text{FN} = 28\cdot 72 \text{ feet.}$$

$$36 \times 2 + 1 = 73, \text{ and}$$

$$\frac{28}{73} = \cdot 38$$

$$\therefore 28\cdot - \cdot 38 = 27\cdot 62 = \text{FL}.$$

This calculation is so easy that the positions of the side-stakes are found in a few seconds. It should be noticed that in the ratios of the slopes, and inclination of the ground, unity is taken for the base of the side slopes, but for the perpendicular of the rise or fall of the ground.

EXAMPLE.

In an embankment, Fig. 57, given the breadth of the roadway AB = 32 feet; the height CF = 18 feet; the side slopes as 1 : $\frac{2}{3}$, (BI : ID :: 1 : $\frac{2}{3}$); the fall of surface from F to M = 1 in 26½, (FN : NM :: 26½ : 1); the rise of surface from F to K, to be the same as the fall, which is very often the case; required the horizontal distances FN and FL, where the surface of the ground will meet the rise slopes of the road.

Fig. 57.

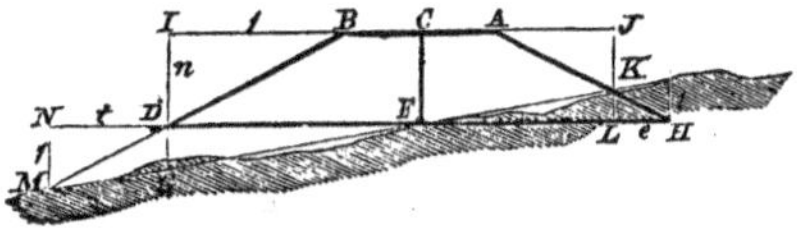

$$\text{DF} = \frac{32}{2} + \frac{3}{2} \times 18 = 43 = \text{FH}.$$

$$26\tfrac{1}{2} \times \tfrac{2}{3} = 17\tfrac{2}{3}$$

Take 1

$$16\tfrac{2}{3}$$

$$\frac{43}{16\frac{2}{3}} = \frac{129}{50} = 2\cdot 58;$$

```
 43·
 2·58
 ----
45·58 = FN.
```

Again:

$$\frac{43}{18\frac{2}{3}} = \frac{129}{56} = 2{\cdot}3, \text{ and}$$

$$43{\cdot} - 2{\cdot}3 = 40{\cdot}7 = \text{FL}.$$

With embankments, the rise answers to the fall in cuttings, by inverting Fig. 53 it becomes Fig. 54, and renders this remark plain. Hence the rule given for cuttings is easily made to answer for embankments.

EXAMPLE.

In a cutting, Fig. 58, given the breadth of the roadway AB = 28 feet; the height CF = 20 feet; the ratio of the side slopes 1 : ½; the inclination of the surface of the ground, at the cross

Fig. 58.

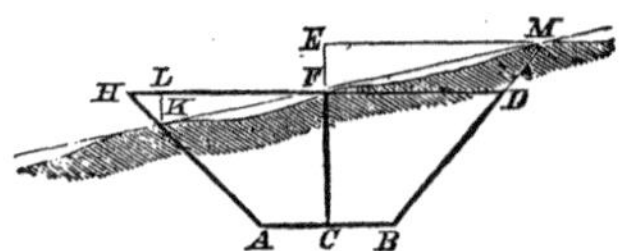

section, taken by a theodolite = 14°; that is, the angle MFD = HFK = 14°. Required the horizontal distances EM, and FL, where the surface of the ground meets the side slopes.

$$90° - 14° = 76°.$$

Then the natural tangent of 76° = 4·010781. Hence the rise and fall of the slope of the surface of the ground at the cross section may be taken as

4·01 to 1.

$$\text{HF} = \frac{28}{2} + \frac{2}{1} \times 20 = 54 = \text{FD}.$$

$$4{\cdot}01 \times \tfrac{1}{2} = 2{\cdot}005$$

Subtract.... 1·000

1·005

$$\frac{54}{1{\cdot}005} = 53{\cdot}73$$

54·00 add half breadth.

EM = 107·73 feet.

Again,

$$4{\cdot}01 \times \tfrac{1}{2} = 2{\cdot}005$$

Add........ 1·000

3·005

$$\begin{array}{rl} & 54 \cdot 00 \text{ from half breadth} \\ \dfrac{\cdot 54}{3 \cdot 005} = & \underline{11 \cdot 31} \text{ take} \\ FL = & 42 \cdot 69 = \text{feet.} \end{array}$$

EXAMPLE.

Fig. 59.

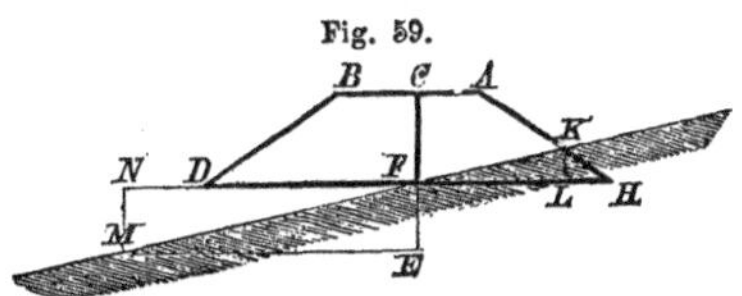

In an embankment, Fig. 56, given the breadth of the roadway AB = 30 ft. the height CF = 12 feet; the ratio of the side slopes 1 : $\frac{3}{4}$; the elevation or inclination of the surface of the ground in the direction of the cross section = 3° (angle KFH = DFM = 3°); required the horizontal distances EM, FL, where the surface of the ground meets the side slopes

$$90^\circ - 3^\circ = 87^\circ$$

$$\tan. 87^\circ = 19 \cdot 081137.$$

Hence the rise and fall of the surface from the centre stake F may be represented by the ratio

$$19 \cdot 08 : 1$$

$$HF = \frac{30}{2} + \frac{4}{3} \times 12 = 31 = FD.$$

$$\begin{array}{rl} 19 \cdot 08 \times \frac{3}{4} = & 14 \cdot 31 \text{ product} \\ & \underline{1 \cdot 00} \text{ subtract} \\ & \underline{13 \cdot 31} \\ \dfrac{31}{13 \cdot 31} = & 2 \cdot 33 \\ & \underline{31 \cdot 00} = FD \\ & \underline{33 \cdot 33} = EM. \end{array}$$

Again,

$$\begin{array}{rl} 19 \cdot 08 \times \frac{3}{4} = & 14 \cdot 31 \text{ product} \\ & \underline{1 \cdot 00} \text{ add} \\ & \underline{15 \cdot 31} \\ & 31 \cdot 00 = FH \\ \dfrac{31}{15 \cdot 31} = & \underline{2 \cdot 03} \text{ subtract} \\ & \underline{28 \cdot 97} = FL. \end{array}$$

The simplicity of this plan of putting down side-stakes is apparent, and the formula from which it is deduced is easily remembered

$$HL = v = \frac{d}{ne + 1}; \ d = FH.$$

$$DN = z = \frac{d}{nt - 1}; \ d = FD.$$

In recapitulating I may as well say that in the ratio

$$1 : n$$

1 represents the base, and n the perpendicular of the side slopes.

In the ratio $t : 1$

t represents the base, and 1 the perpendicular of a rise, in the case of a cutting, and *vice versa* in the case of an embankment. See Figs. 57 and 58.

In the ratio $e : 1$

e represents the base, and 1 the perpendicular.

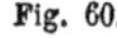

PROBLEM.

To set out the Widths when the surface of the ground is laterally sloping, and when the cross section consists partly of Cutting and partly of Embanking.

Fig. 60.

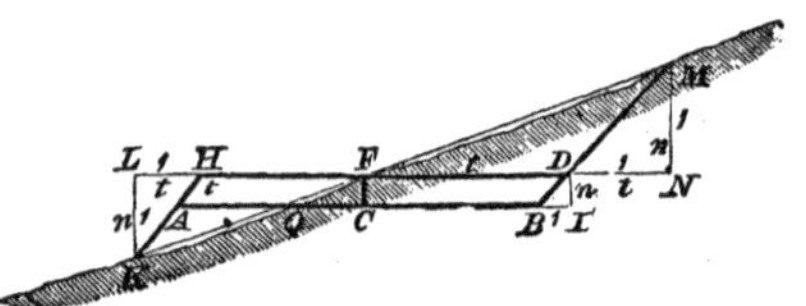

Let AKQMB be the cross section of a railroad, consisting partly of a cutting QMB, and partly of an embankment AQK. Let b = AB = HD, the bottom width, F the centre stump, FC = h, the depth of the cutting; KM the sloping surface of the ground. First, suppose the embankment AKQ to occupy less than half the bottom width, as from A to Q. And further suppose the ratio of the slope of the earth on both sides of the centre stake F, to be represented by the same ratio,

$$t : 1,$$

that is, FN : NM :: t : 1.
or FL : LK :: t : 1.

Yet the reasoning upon which the practical rule is founded will apply to two different slopes of ground, from the centre stake F, as in the former case. When the slope of the ground is known, or the ratio $t : 1$ is determined, then the point Q, where the slope meets the base, is readily found, for,

$$1 : t :: \text{FC} : \text{CQ}$$
$$\text{or CQ} = th.$$

that is if the height h, multiplied by t be less than AC, half the breadth of the roadway, the work will be at this cross-section part cutting and part embanking.

Let the side slopes be represented by the ratio $1 : n$, that is,

$$\text{BI} : \text{ID} = \text{DN} : \text{NM} = \text{HL} : \text{LK} = 1 : n.$$
$$\text{Put DN} = x \text{ and HL} = y.$$

The horizontal half breadth FD, through F, to meet the slope of the cutting, is found as in the former investigation, where the slope of the ground does not meet the road-way at either side of F.

$$\text{FD} = \frac{b}{2} + \frac{1}{n} \times h, \text{ which put} = d.$$

The horizontal half breadth FH, through F, to meet the slope of the embankment KH is easily found when FD is known, for HD = AB = b, and HF = HD — FD =

$$b - \left(\tfrac{1}{2} b + \frac{1}{n} \times h\right). =$$
$$\frac{b}{2} - \frac{1}{n} h, \text{which put} = \delta = \text{HF}.$$
$$1 : n :: x : nx = \text{NM};$$
$$\text{Also, } t : 1 :: d + x : \frac{d + x}{t} = \text{NM};$$
$$\therefore nx = \frac{d + x}{t}, \text{ and}$$
$$ntx - x = d,$$
$$\therefore x = \frac{d}{nt - 1};$$

this result is exactly the same as that given before, for the increase of horizontal breadth in the case of a rise in cutting or a fall in embanking.

$$1 : n :: y : ny = \text{KL};$$
$$\text{again, } t : 1 :: \delta + y : \frac{\delta + y}{t} = \text{KL};$$
$$\therefore ny = \frac{\delta + y}{t}$$
$$\text{and } y = \frac{\delta}{nt - 1},$$

which is a similar expression to that given for x; the only difference is that $\delta =$ HF is put in the place of $d =$ FD.

EXAMPLE.

Let AB $= b = 28$ feet; FC $= h = 4$ feet; the side slopes 1 : 2 (BT : ID : : 1 : 2); the cross slope of the ground 13 : 1, (FN : NM : : 13 : 1); required the horizontal distances FN and FL where the slopes meet the surface.

$$\text{FD} = \frac{28}{2} + \tfrac{1}{2} \times 4 = 16 = d.$$

$$\text{FH} = 28 - 16 = 12 = \delta.$$

Now $n = 2$, and $t = 13$;

$$\therefore \frac{16}{2 \times 13 - 1} = {\cdot}64 \text{ and } \frac{12}{2 \times 13 - 1} = {\cdot}48.$$

Consequently $16{\cdot}64 =$ FN and $12{\cdot}48 =$ FL.

Fig. 61.

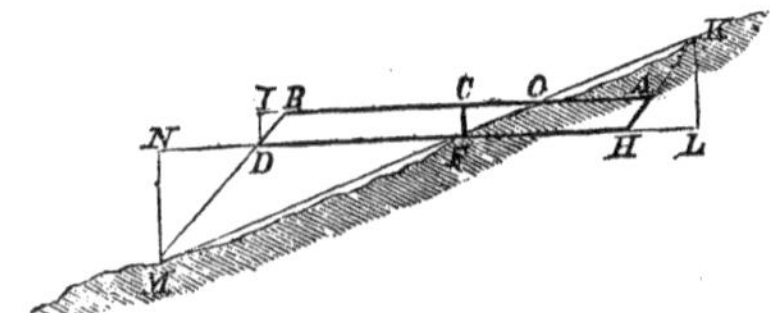

EXAMPLE.

In Fig. 61, the embankment MBQ occupies more than half the bottom width, as from B to Q; let AC = CB = 14 feet; FC = 4 feet.

FL : LK : : 13 : 1;
HL : LK : : 1 : 2;

as in the last example, required the horizontal distances from F to L and from F to N.

$$\text{FD} = \frac{28}{2} + \frac{1}{2} \times 4 = 16 = d$$

$$\text{FH} = 28 - 16 = 12 = \delta.$$

Now, $n = 2$, and $t = 13$.

$$\therefore \frac{16}{2 \times 13 - 1} = {\cdot}64; \text{ and } \frac{12}{2 \times 13 - 1} = {\cdot}48.$$

Hence FN $= 16{\cdot}64$, and FL $= 12{\cdot}48$ feet.

EXAMPLE.

In Fig. 62, the embankment AQK occupies less than half the bottom width from Q to A; let AC = CB = 14 feet; FC = 4 ft.;

FL : LK :: 13 : 1;
BI : ID :: 1 : 2;

Required the horizontal distances from F to L, and from F to N.

Fig. 62.

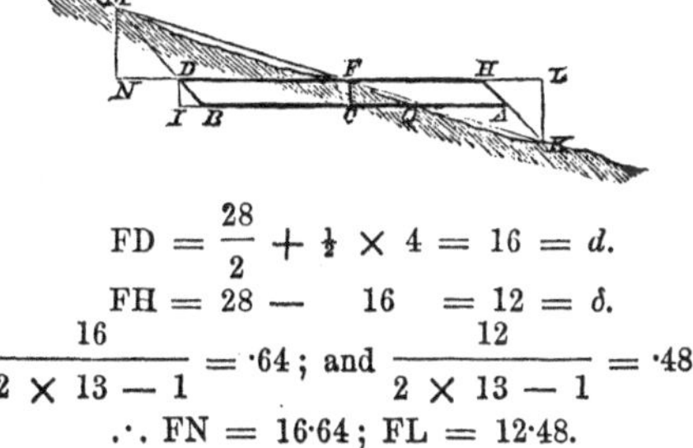

$$FD = \frac{28}{2} + \tfrac{1}{2} \times 4 = 16 = d.$$

$$FH = 28 - 16 = 12 = \delta.$$

$$\frac{16}{2 \times 13 - 1} = \cdot 64 \text{; and } \frac{12}{2 \times 13 - 1} = \cdot 48$$

$$\therefore FN = 16 \cdot 64 \text{; } FL = 12 \cdot 48.$$

EXAMPLE.

In Fig. 63, the embankment QBM occupies more than half the bottom width, from Q to B. Let AC = CB = 14 ft.; FC = 4 ft.;

FL : LK :: 13 : 1;
HL : LK :: 1 : 2;

Required the horizontal distances from F to L and from F to N.

Fig. 63.

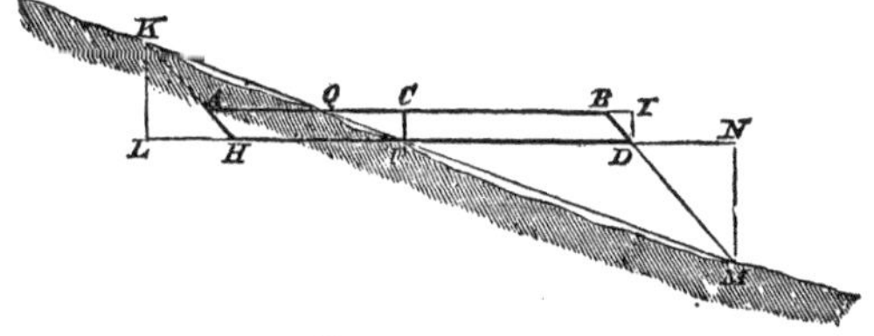

$$FD = \frac{28}{2} + \tfrac{1}{2} \times 4 = 16 = d.$$

$$FH = 28 - 16 = 12 = \delta.$$

$$\left.\begin{aligned} \mathrm{DN} &= x = \frac{d}{nt-1} = \cdot 64 \\ \mathrm{HL} &= y = \frac{\delta}{nt-1} = \cdot 48 \end{aligned}\right\} \text{as in the foregoing examples;}$$

and FN = 16·64 ; FL = 12·48.

By comparing the figures 60, 61, 62, and 63, and observing how the formulas

$$x = \frac{d}{nt-1} = \mathrm{DN},$$

$$\text{and } y = \frac{\delta}{nt-1} = \mathrm{HL},$$

may be applied to every possible case, the setting out of side-stakes, when part of the cross section is a cutting and part an embankment, becomes easy.

METHODS OF FINDING THE CONTENTS

OF

RAILROAD CUTTINGS & EMBANKMENTS.

To find the contents of railway cuttings and embankments with accuracy is one of the most important problems in railroad engineering practice.

The first object to be acquired in preparing the dimensions of the different cross sections is to reduce the irregular sections to a level, so that the level section may have the same area as the irregular one.

Fig. 64.

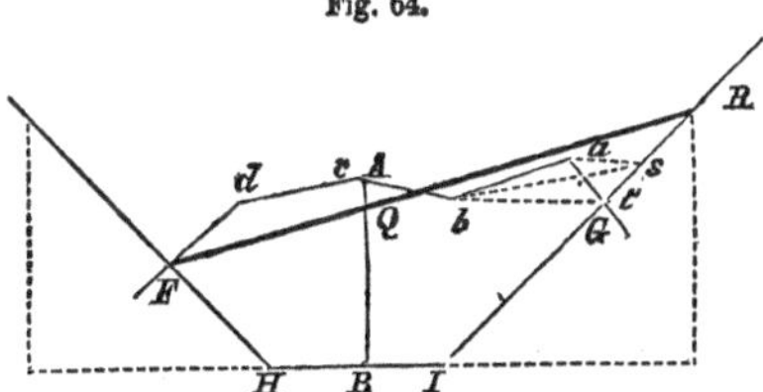

To draw a line FR, so that the area of any cross section IG*abcd* FH may be equal to the area of the figure HFRI requires but little engineering or geometrical skill; a thread applied across the boundary, when the inequalities are not great, will, in most cases, be sufficient to determine FR. When the variations are great, the parallel-ruler will determine FR, in a few seconds, with mathematical accuracy.

Thus draw *as*, parallel to *b*G, and then draw *sb*; then the triangle *bs*G = *ba*G; through *b* draw a line parallel to *cs*, join *tc*, as in the last case, and continue the process, until the line FR is determined.

The cross section is easily reduced to the form FBCE, Fig. 65; but to draw IH so that it may be parallel to EF, and that the area

$$\text{IHFE} = \text{FBCE},$$

has employed the ingenuity of engineers without much accurate practical success, although the object may be thus easily obtained by construction.

Fig. 65.

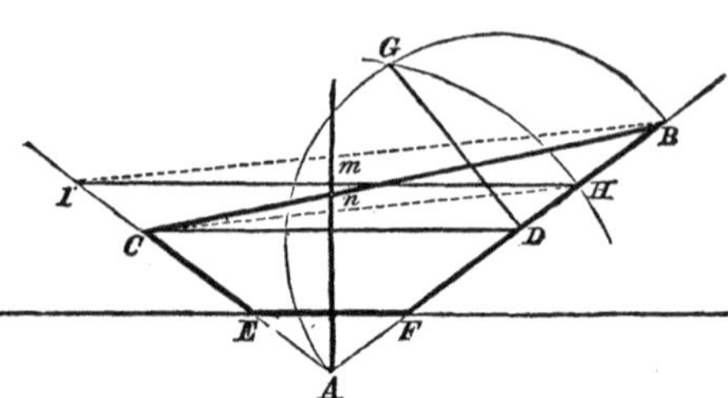

On AB describe a semicircle; draw CD parallel to EF, and DG perpendicular to AB; with AG describe the arc GH, through H draw HI parallel to CD; then the area of IHEF = area of BCEF, and *Am* is the height of the equivalent level cutting required. The same construction will suit for embankments. When this construction is performed accurately, then IB is parallel to CH. Hence the truth of the construction may be tested by taking a parallel-ruler and finding whether the lines IB and CH are parallel or not; if these lines be not parallel, the construction must be repeated.

Fig. 66.

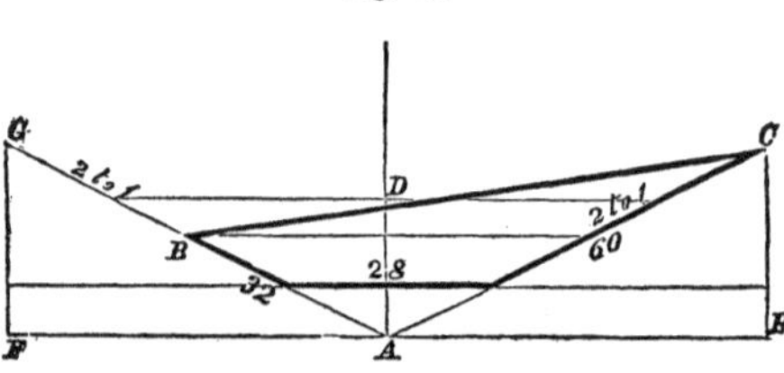

EXAMPLE.

Let AB, Fig. 66, be = 32 feet; AC = 60 feet; slopes 2 : 1 (AE : EC :: 2 : 1); required the length of AD.

$$\frac{60 \times 32}{5} = 384 = \text{the spuare of AD};$$

therefore,

$$\text{AD} = \sqrt{384} = 19{\cdot}596 \text{ feet}$$

$1^2 + 2^2 = 5$, the number above divided by.

A rule much easier than this cannot be expected. I will apply it to another example in which the ratio of the slopes is expressed in more compound numbers.

EXAMPLE.

Given AB (Fig. 66) = 26 feet; AC = 50 feet; slopes $1\frac{1}{2}$ to 1 (AF : FG :: $1\frac{1}{2}$: 1); required AD.

$$\frac{3^2}{2} + 1^2 = \frac{13}{4}$$

$$\frac{26 \times 50 \times 4}{13} = 400, \text{ the square of AD},$$

$$\therefore \text{AD} = \sqrt{400} = 20.$$

It generally happens that AB, the depth in or over the centre A, the breadth of the roadway EF, and the slopes of *fa*, F*a*, *f*E are only given, and from which the depth DA of the equivalent level cutting is required. The point B may or not be in the surface of the ground. For example, in Fig. 64, the point A represents the centre stake, and is in the surface of the ground, while Q, where FR meets AB, is below the surface of the ground; hence, the point B, in Fig. 67, may be above or below the surface, as the case may

Fig. 67.

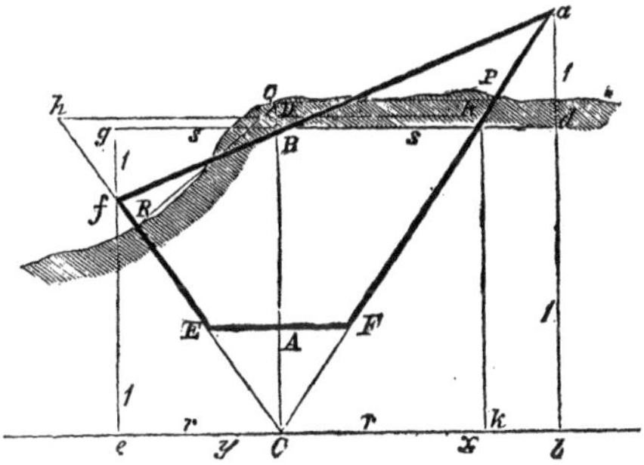

be. This remark is important, as the position of the centre stake is so much referred to in setting out the side slopes; and, in fact, it is the point from which all measures are taken. When the height of the centre stake is known above or below the middle of the roadway, in cutting or embanking, the position of B in the line *fa*, that equalizes the surface, is not far from the centre stake in most cases. However, *fa* is the line found by construction or otherwise that makes the

$$\text{Area EF}af = \text{area EFPQR}.$$

Q being the centre stake, and QR and QP, Fig. 67, the lines that are used to find the places of the side-stakes R and P.

Let the slope of fa be represented by $s : 1$, and of Cf or Ca by r to 1. Also let BC be represented by m, then, putting $x = Cb$ and $y = Ce$, by similar triangles,

$$r : 1 :: x : \frac{x}{r} = ab$$

$$s : 1 :: x : \frac{x}{s} = ad$$

$$\therefore m + \frac{x}{s} = \frac{x}{r}, \text{ and hence } x = \frac{rsm}{s-r} = Cb.$$

Again,

$$r : 1 :: y : \frac{y}{r} = ef;$$

$$s : 1 :: y : \frac{y}{s} = gf;$$

$$\therefore m - \frac{y}{s} = \frac{y}{r}, \text{ and } y = \frac{rsm}{s+r} = Ce.$$

$$\frac{y}{r} = ef = \frac{sm}{s-r};$$

$$\frac{x}{r} = ab = \frac{sm}{s-r};$$

But on referring to Fig. 67 it will be seen that

$$Ca^2 = Cb^2 + ab^2;$$

$$\text{and } Cf^2 = Ce^2 + ef^2;$$

$$\therefore Ca = \sqrt{\frac{s^2 m^2 + r^2 s^2 m^2}{(s-r)^2}};$$

$$Cf = \sqrt{\frac{s^2 m^2 + r^2 s^2 m^2}{(s+r)^2}};$$

But because the areas of the triangles Cfa and Chh are equal, and the side (Fig. 67.)

$$Ch = Ch$$

Then,

$$Ch = \sqrt{Cf \times Ca};$$

$$\therefore Ch = \sqrt[4]{\frac{s^2 m^2 (1 + r^2) \times s^2 m^2 (1 + r^2)}{(s+r)^2 \times (s-r)^2}} =$$

$$\frac{\sqrt{s^2 m^2 (1 + r^2)}}{(s+r)(s-r)}.$$

But

$$Ch : CD\ (hk) :: \sqrt{(1 + r^2)} : 1.$$

Hence,

$$CD = \frac{sm}{\sqrt{(s+r)(s-r)}} =$$

the depth of the equivalent level cutting. This computation is easily effected by logarithms, the use of which are recommended when the numbers expressing the ratios of the slopes are compound.

RULE BY LOGARITHMS.

From the sum of the logarithms of s and m, take half the sum of the logarithms of $s + r$ and $s - r$, and the remainder is the logarithm of CD (Fig. 67), the depth of the equivalent cutting where the side slopes meet.

This rule is given here for the first time, and is the only simple and, at the same time, exact rule that has appeared.

EXAMPLE.

Let the roadway EF = 28 feet wide, with the side slopes of $1\frac{1}{2} : 1$ and the ground to incline transversely at an angle of 15° ; with a depth BA, at the station B, of 20 feet; required AD the depth of the equivalent level cutting FE *hh*. See Fig. 68.

Fig. 68.

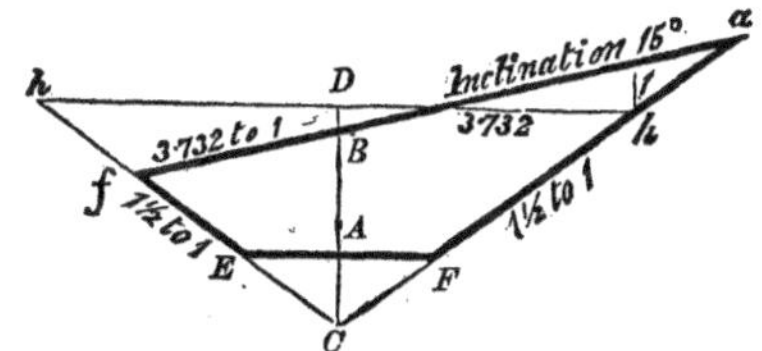

Since the cotangent of 15° = 3·732, the inclination of *fa* is 3·732 to 1; hence $s = 3·732$; $r = 1·5$.

$$1·5 : 1 :: 14 : 9\tfrac{1}{3} = AC.$$

$$\therefore BC = 20 + 9\tfrac{1}{3} = 29·333$$

	3·732		3·732
	1.500		1·500
	2·232 = $s - r$;		5·232 = $s - r$;
Log. 2·232 =	·3486942	Log. 3·732 =	·5719416
log. 5·232 =	·7186677	log. 29.333 =	1·4673565
	2)1·0673619	From	2·0392981
	·5336809	Take	·5336809

CD = 32·034 logarithm = 1·5056172
9·333 = AC

22·701 = AD, required.

To find AD by common arithmetical calculation is not very difficult, especially when a table of squares, &c., of numbers is employed.

The general expression may be made to assume the form.

$$CD = \frac{sm}{\sqrt{s^2 - r^2}}.$$

EXAMPLE.

Let the road be 36 feet wide, with side slopes of 2 : 1, the ground to incline transversely at a slope of 4 : 1; and the depth over the centre of the road, AB = 24 feet; what is the depth of an equivalent level cutting?

$$2 : 1 :: \frac{36}{2} : 9 = CA.$$

$$\begin{aligned} 24 &= AB; \\ 9 &= AC; \\ m = 33 &= BC. \\ m = 33 & \\ s = 4 & \\ 132 &= m \times s. \end{aligned}$$

$$\sqrt{s^2 - r^2} = \sqrt{4^2 - 2^2} = \sqrt{12},$$

the square root of 12 can be found in the table = 3·464, which is taken sufficiently near for the purpose.

$$\frac{132}{3{\cdot}464} = 38{\cdot}103 = CD.$$

$$\begin{aligned} 9{\cdot}000 &= CA. \\ 29{\cdot}103 &= AD. \end{aligned}$$

EXAMPLE.

The breadth of the roadway is 25 feet, with side slopes of 2½ : 1; and a transverse ground slope of 6 : 1 (that is, the slope that equalizes the cross sectional area); the depth of the station from the centre of the road = 15 feet, what is the depth of a level cutting of equal area?

$$2\tfrac{1}{2} : 1 :: \frac{25}{2} : 5 = AC.$$

$$\begin{aligned} 15 + 5 = 20 &= m. \\ 6 &= s. \\ 120 &= m \times s. \end{aligned}$$

$$\begin{aligned} 6^2 &= 36 \\ (2\tfrac{1}{2})^2 &= 6{\cdot}25 \\ \sqrt{29{\cdot}75} &= 5{\cdot}45. \end{aligned}$$

The square root of such numbers as 29·75 can be found by inspecting the table of squares, square roots, &c.

$$29{\cdot}75 \times 4 = 119.$$

The square root of 119 from the table = 10·9087121, the half of which = 5·454356, of which we have taken 5·45.

$$\frac{120}{5{\cdot}45} = 22{\cdot}02 = CD.$$

$$\underline{5{\cdot}00} = AC.$$

$$\underline{17{\cdot}02} = AD.$$

This calculation would be very concise, only the work is accompanied by explanations. Without such rendering the work might stand thus, as half the square root of 119 can be taken from the table without calculation.

$$\frac{25}{5} = 5 \qquad 6^2 - (\tfrac{5}{2})^2 = \frac{119}{4}$$

```
      5
     15
     --
     20
      6
    ----
5·45) 12000        { 22·02
      1090         {  5·00
      ----           -----
       1100          17·02 = AD.
       1090
       ----
        1000
        1090
        ----
```

EXAMPLE.

The breadth of the roadway is 27·8 feet, with side slopes of 36° 40′ and a transverse equalizing ground slope of 21° 15′; the depth of the station from the centre of the road 45·6 ft.; what is the depth of a level cutting of equal area?

Natural cotangent of 36° 40′ = 1·3432 = r
Natural cotangent of 21° 15′ = 2·5715 = s

$$s + r = 3{\cdot}9147$$

$$s - r = 1{\cdot}2283$$

$$2r = \frac{27.8}{2{\cdot}6864} = 10{\cdot}35 = AC.$$

$$\begin{array}{r} 45{\cdot}6 \\ 10{\cdot}35 \\ \hline 55{\cdot}95 \end{array} = m$$

Log. $r + s$ =	·5926985	Log. m =	1·7478001
log. $r - s$ =	·0893045	log. s =	0·4101865
	2)·6820030	From	2·1579866
	·3410015	Take	·3410015
		log. 65·612 =	1·8169851

65·612
10·350

55·262 = AD,

the depth of a level cutting of equal area with the one defined in the question.

EXAMPLE.

The breadth of the roadway = 33·7 ft ; the side slopes of an embankment are 19 : 7 ; and a traverse ground slope of 23 : 3 ; the depth of the embankment from the centre of the road = 18·4 feet ; what is the depth of the level embankment of equal cross area.

$$s = \frac{23}{3} = 7.6667$$

$$r = \frac{19}{7} \quad 2{\cdot}7143$$

$$10{\cdot}3810 = s + r$$

$$4{\cdot}9524 = s - r$$

$$2r = \frac{33{\cdot}7}{5{\cdot}4286} = 6{\cdot}2078$$

$$18{\cdot}4000$$

$$24{\cdot}6078 = m.$$

Log. $s + r$ =	1·0162392	Log. m =	1·3910728
log. $s - r$ =	·6948157	log. s =	0·8846085
	2) 1·7110549	From	2·2756813
	·8555274	Take	·8555274
		Logarithm of 26·312 =	1·4201539

26·312
6·2078

AD = 20·1042 feet.

CONSTRUCTION.

When many constructions are to be made, section or cross-barred paper will be found very convenient. This paper is ruled, or rather printed from plates of steel or copper, to suit with great accuracy a variety of scales.

Take AC = 18·4 feet and produce it both ways; draw EAF = 33·7 feet and perpendicular to AC; AE = AF.

Produce AF to n and make Fn = 19 on any convenient scale of equal parts, draw mn perpendicular to Fn and = 7 such parts, then draw BFmH, and in the same way draw BEGL. Again make Ct = 3, and tv perpendicular to it = 23, draw v GCH, and EFGH is the cross section of the embankment. Draw GI parallel to EF,

Fig. 69.

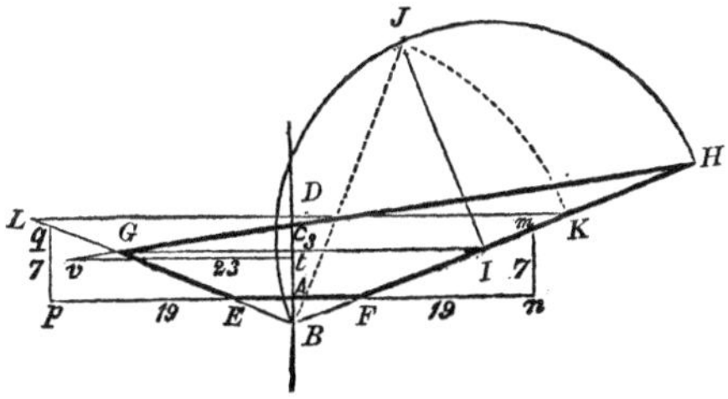

on BH describe the semicircle BJH, draw IJ perpendicular to BH, and make BK = BJ, draw KL parallel to EF, then the area FEGH = the level area LKFE and

AD = 20·1 feet.

as before found by calculation.

PROBLEM.

To find the Contents of Cutting and Embankments.

Let m, be the breadth of the bottom of a level cutting at the rails;

$$a, b, c, d, \ldots\ldots\ldots z,$$

the perpendicular heights taken n feet apart; and $r : 1$ the ratio of the slopes, then,

The content of the central part

$$= \frac{mn}{2}\left\{a + z + 2\,(b + c + d\ldots)\right\};$$

where a and z are the extreme ordinates and $b, c, d, e\ldots.$ the intermediate.

The content of the two slopes =

$$\frac{nr}{3}\left\{a + b)^2 + (b + c)^2 + (c + d)^2\ldots\ldots [ab + bc + cd\ldots.]\right\}.$$

EXAMPLE.

Required the solid content of a cutting or embankment ABCDEFGH whose heights taken at 1 chain of 66 feet apart are 30 feet $= a$, and 20 feet $= b$; the width of the rails $=$ 36 feet $= m$; the slopes 2 to 1.

In this example $r : 1$ becomes $2 : 1$; $n = 66$ feet.

The general formula becomes

$$\frac{mn}{2}\left\{a + b\right\} + \frac{nr}{3}\left\{(a + b)^2 - ab\right\},$$

when two ordinates, a and b are used.

$$\begin{array}{rr} & 20 \\ & \underline{30} \\ & 50 \\ & \underline{66} \\ & 3300 \\ & \underline{36} \\ & 19800 \\ & \underline{9900} \\ & 2)\underline{118800} \\ \text{ABPNKFGI} = & \underline{59400} \text{ cubic feet,} \end{array}$$

$$\begin{array}{rrl} a + b = & 50 & \\ & \underline{50} & = a + b \\ & 2500 & \\ 30 \times 20 = & \underline{600} & = ab \\ & 1900 & \\ & \underline{2} & = r \\ & 3800 & \\ & \underline{66} & = n \\ & 22800 & \\ & \underline{22800} & \\ & 3)\underline{250800} & \\ & \underline{83600} & = \text{cubic feet} \end{array}$$

in PBCHIG and ANDEKF together.

$$\begin{array}{rl} 83600 & \\ \underline{59400} & \\ \underline{143000} & = \text{whole content} \end{array}$$

Fig. 70.

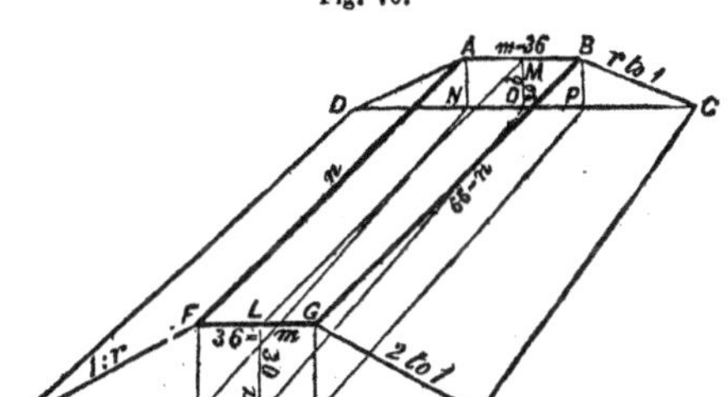

in cubic feet. This number divided by 27, gives 5296·296, the content in cubic yards. I have used no help to contract the process in order that the nature of the problem may be thoroughly understood.

EXAMPLE.

Let the cubical contents of the cutting or embankment mentioned in the last example be required, when $n = 100$ feet.

$$
\begin{array}{rl}
a = & 20 \\
b = & \underline{30} \\
2) & \underline{50} \\
 & 25 \times 100 = 2500 = \text{area.}
\end{array}
$$

$$
\begin{array}{r}
2500 \\
\underline{36} \\
15000 \\
\underline{7500} \\
90000
\end{array}
= \text{cubic feet in ABPNKFGI.}
$$

$$
\begin{array}{rrl}
 & 50 & = a + b \\
 & \underline{50} & \\
 & 2500 & = (a + b)^2 \\
ab = & \underline{600} & \\
 & 1900 & \\
 & \underline{2} & = r \\
 & 3800 & \\
 & \underline{100} & = n \\
3) & 380000 & = 126666{\cdot}66 \text{ cubic feet,}
\end{array}
$$

the content of the two slopes.

```
        90000
       126666·66
27)    216666·66
         8024·7  whole content in cubic yards.
```

Now observe how easily this result can be obtained when the former 5296·296 is found for a chain of 66 feet.

```
                          5296·296
                 half  =  2648·148
                        ( 7944·444  for 99 feet
Passing two figures  }      79·444
  to left each time  }        ·794
                        (        7
                          8024·689
```

Consequently, if any cubic yards as 90000 be for a 66-feet chain,

```
 90000
 45000
135000
  1350
    ·13·5
       ·1
136363·6
```

will be the cubic yards for 100 feet. This result is obtained without mental labor.

EXAMPLE.

Required the cubical content of these cuttings by inspecting the following tables I., II., and III.

First for n = 66 feet.

```
    20
    30
2)  50
    25 × 36 = 900 feet.
```

From the small table (I.), for an area of 900 square feet, there is given 2199·9999 cubic yards, which may be taken as 2200 cubic yards.

In the large table (III.), over 20 and opposite 30 will be found

```
1548
   2 = r
3096
2200
5296 cubic yards,
```

TABLE I.—*For 66 Feet.*

1	2·4444444
2	4·8888888
3	7·3333333
4	9·7777777
5	12·2222222
6	14·6666666
7	17·1111111
8	19·5555555
9	21·9999999

the same whole number of yards as these before found, according to the formula.

When the cubic content in yards for a chain of 66 is known, the content for a chain of 100 is found instantly as before shown.

```
5296
2648
────
7944·   for 99 feet.
  79·44
    ·79
───────
8024·23 for 100 feet.
```

When the work is not encumbered by explanations, the ease of application is very apparent as the following examples will show.

EXAMPLE.

Let the equivalent level cutting at one end have a particular height JL = 25 feet (Fig. 70); at the other end, OM = 20. What is the cubical content for lengths of 100 feet, 66 feet, and 50 feet, the roadway 28 feet wide and the side slopes $1\frac{1}{2}$ to 1.

From Table I.

```
 25
 20
 ──
 45
 14 = half 28.
───
180
45
───
630
```

For 600,	1466·6666
30,	73·3366
	1540·

Table III.

Opposite 25 under 20,	1343· } $r = 1\frac{1}{2}$
	621· }
	1864
	1540
For a length of 66 ft.,	3404 cubic yds.

	3404
	1702
For 99 ft.	5106
	51·06
	51
For 100 ft.	5157·57 cubic yds.

2)5157·57 for 100 feet.
2578·78 for 50 feet.

Hence, if 25 and 20 be the heights of a filling or the depths of a cutting,

	Cubic yds.
For 66 feet, the content =	3404·
for 100 feet, the content =	5157·5
for 50 feet, the content =	2578·8

It is easily seen by a practical Civil Engineer that this is the best and easiest method of finding the solid content of cuttings or embankments yet proposed, whether the chain be 100 feet long, 66 feet, or 50.

The content for any other distance, as 121·3 feet is also readily determined, thus :

for 100 feet,	5157·5
for 1 foot,	51·575
	121·3
	154725
	51575
	103150
	51575
for 121·3 ft.,	6256·0475

Any other length may be applied in the same manner.

EXAMPLE.

Let 16 feet be the height of a level filling, which have the same area as the cross section at this station; 100 feet from this the height of the level filling is found to be 14 feet; how many cubic yards of earth does it contain, the ratio of the side slopes ¾ to 1, breadth of the roadway = 32 feet.

TABLE II.—*For* 100 *Feet.*

1	3·703703
2	7·407407
3	11·111111
4	14·814814
5	18·518518
6	22·222222
7	25·925925
8	29·629629
9	33·333333

From Table II., which is for finding the content of the central part for lengths of 100 feet, will be found the content of the central part, thus.

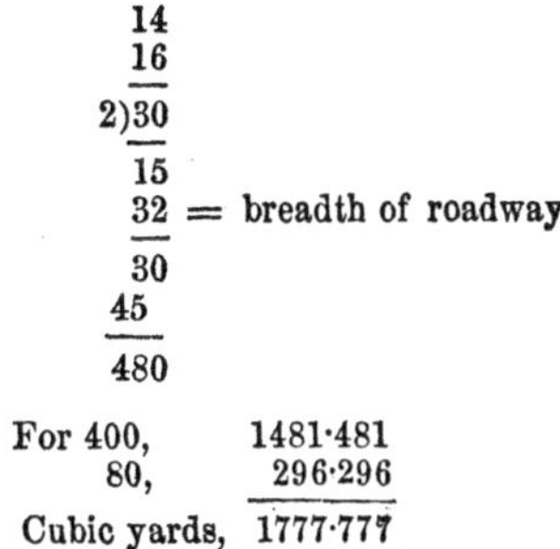

14
16
2)30
15
32 = breadth of roadway.
30
45
480

For 400, 1481·481
80, 296·296
Cubic yards, 1777·777

Table III.

For the content of the two slopes and a length of 66 feet.

Opposite 16 over 14, 551
$\frac{3}{4} = r.$
413·25

413·25
half = 206·625
619·875
6·198
61
626·134
Cubic yards, 1777·777 for 100 feet.
2403·911 whole content.

EXAMPLE.

Suppose the equivalent level cutting at one end to be 24 feet, and at the other 36; the roadway 31 feet wide; the length of the cutting 100 feet, the side slopes $1\frac{1}{3}$ to 1; required the cubical content in cubic yards.

24
36
2) 60
30
31 breadth of roadway.
930

From Table II.

900.................. 3333·333
30 .... 111·111
3444·444

From Table III.

Opposite 36 over 24.... 2229
4
$r = \frac{4}{3}$ 3) 8916
For 66 feet 2972
half.............. 1486
4458
44·58
·45
For 100 feet .... 4503·03
3444·44
Total content = 8947·47

EXAMPLE.

Let a cutting be in every respect the same as the last, only the length = 66 feet; what is the cubical content?

```
36 + 24 = 60
          31  = breadth of roadway.
          60
         180
      2) 1860
          930

      From Table I.
For 900............ 2200
     30............   73·33
                    2273·33

Opposite 30 over 24 .. ....... 2229
                                  4
               r = 4/3   3)  8916
                             2972
                             2273⅓
Total content for 66 feet    5245⅓
```

As all the figures employed are set down, it is evident that this plan is superior to any other proposed method as but one-tenth the mental labor is expended.

What is the content for 12·3 feet when the content for a length of 66 ft. = 5245·33 cubic feet.

```
           5245·33
           2622·66
           7868·00
             78·68
             78
           7947·47 for 100 feet.
           79·4747 for 1 foot.

           79·4747
              12·3
           2284241
          1589494
          794747
Cubic feet 876·53881 for 12·3 feet length.
```

EXAMPLE.

Required the cubical content by the general formula.

$$\frac{m\,n}{2}\,(a + b) + \frac{nr}{3}\,(a^2 + ab + b^2).$$

$$a = 24;\ b = 36$$
$$r = 1\tfrac{1}{3};\ m = 31;\ n = 100 \text{ feet.}$$

$$\frac{31 \times 100}{2} \times (24 + 36) = 93000$$

```
                  24
                  36
                  --
(a + b)²   =     60²  =   3600
 a × b     =               864
                          ----
a² + ab + b²   =          2736
                             4
                         -----
                     3)  10944
                         -----
                          3648
                           100
                        ------
                   3)   364800
                        ------
                        121600
                         93000
                        ------
Content in cubic feet = 214600
```

```
27)214600(7948.
   189
   ---
    256
    243
    ---
     130
     108
     ---
      220
      216
      ---
```

The difference existing between the result obtained by using the tables and the result given by the formula is but very small for so large a content.

It arises from the results given in Table III. being whole numbers without decimals. In fact, Table III. contains the content in cubical yards to the nearest unit, and the depths of the level cuttings have all the integral values from 1 to 70.

When decimals are annexed, the additional cubical content is found by consulting Table IV. The method of using this table will best appear from example.

EXAMPLE.

Let $a = 52{\cdot}6$ feet; $b = 30{\cdot}4$; the slopes 2 : 1; the breadth of the roadway 36 feet; length 66 feet: what is the content in cubic yards?

$$\begin{array}{r l} 52{\cdot}6 & \\ 30{\cdot}4 & \\ \hline 83{\cdot}0 & \\ 18 & = \text{half } 36. \\ \hline 664 & \\ 83 & \\ \hline 1494. & \end{array}$$

From Table I.

For 1000,........................	2444·444
" 400,..................... ...	977·777
" 90,..........	219·999
" 4,.......................	9·777
	3651·997

Table III.

Opposite 52 under 30,................. 4208.

To find what must be added for decimals (·6) and (·4) employ

Table IV.

$$\begin{array}{r l} 52{\cdot}6 & \\ 2 & \\ \hline 105{\cdot}2 & = 2a \\ 30{\cdot}4 & = b \\ \hline 135{\cdot}6 & = \\ 13{\cdot}56 & = \dfrac{2b + a}{10} \end{array}$$

The nearest whole number to which is 14.

Opposite 14 and under ·6, ... 68·

$$\begin{array}{r l} 30{\cdot}4 & = b \\ 2 & \\ \hline 60{\cdot}8 & \\ 52{\cdot}6 & = a \\ \hline 11{\cdot}34 & = \dfrac{2b + a}{10}. \end{array}$$

The nearest whole number to which is 11.

TABLE IV.

	·1	·2	·3	·4	·5	·6	·7	·8	·9
1	1	2	2	3	4	5	6	7	7
2	2	3	5	7	8	10	11	13	15
3	2	5	7	10	12	15	17	20	22
4	3	7	10	13	16	20	23	26	29
5	4	8	12	16	20	24	29	33	37
6	5	10	15	20	24	29	34	39	44
7	6	11	17	23	28	34	40	46	51
8	7	13	20	26	33	39	46	52	59
9	7	15	22	29	37	44	51	57	66
10	8	16	25	33	41	49	57	65	73
11	9	18	27	36	45	54	63	72	81
12	10	20	29	39	49	59	69	78	88
13	11	21	32	42	53	63	74	85	95
14	11	23	34	46	57	68	80	91	103
15	12	24	37	49	61	73	86	98	110
16	13	26	39	52	65	78	91	104	117
17	14	28	42	55	69	83	97	111	125
18	15	29	44	59	73	88	103	117	132
19	16	31	47	62	77	93	108	124	139
20	16	33	49	65	82	98	114	130	147
21	17	34	51	68	86	103	120	137	154

Opposite 11 and under ·4,................ 36.

$$\left.\begin{array}{r} 4208 \\ 68 \\ 36 \end{array}\right\} \text{add}$$

$$\begin{array}{r} 4312 \\ 2 \end{array} = r$$

$$\text{add} \left\{\begin{array}{r} 8624 \\ 3652 \end{array}\right.$$

$$\underline{12276} \text{ cubical content}$$

for a length of 66 feet.

If the length = 100 feet with the other dimensions remaining the same, the content will be

$$\begin{array}{r} 12276 \\ 6138 \\ \hline 18414\cdot \\ 184\cdot14 \\ 1\cdot84 \\ \hline \end{array}$$

for 100 ft., 18599·98 cubic yards.

These results do not differ from those obtained with mathematical accuracy more than two yards, according to the formula.

$$\frac{nm}{2}\left\{a + b\right\} + \frac{nr}{3}\left\{(a + b)^2 - ab\right\}$$

$$a = 52{\cdot}6,\ b = 30{\cdot}4.$$

$$m = 36\ ;\ n = 66\ ;\ \text{all in feet};\ r = 2.$$

$$\frac{36 \times 66}{2} \times (52{\cdot}6 + 30{\cdot}4) = 36 \times 33 \times 83 = 98604.$$

```
                  52·6
                  30·4
                  ----
                  83·0
                  83·
                 -----
                 249
                664
                ----
                6889      = (a + b)²
30·4 × 52·6 =   1599·04   = a × b
                -------
                5289·96
                      2   = r
              ---------
              10579·92
                  · 22    = n/3
              ---------
              2115984
              2115984
              ---------
              232758·24   = rn/3 (a² + ab + b²)
               98604·
              ---------
           27)331362·26
              27
              --
               61          { 12273, the true content
               54          { in cubic yards.
               --
                73
                54
                --
                196
                189
                ---
                  72
                  81
```

I will next give a model example, merely setting down the numbers employed in the operation.

EXAMPLE.

Let the depths of a cutting be 39·3 and 37·7 feet; their distance = 66 feet; the ration of the slopes $1\frac{1}{2}$ to 1; what is the content? Bottom width = 33.

```
     39·3
     37·7
     ----
     77·0
     33
    -----
    231
   231
   -----
 2)2541·
   ------
   1270·5
```

Table I.

For 1000, ...	2444·444
" 200,........................	488·888
" 70,........................	171·111
" ·5,........................	1·222
	3105·666

Table III.

Opposite 39 over 37 3531

```
39·3*
   2
----
78·6
37·7
----
```

$$\frac{2a+b}{10} = 11{\cdot}63$$

Table IV.

Opposite 12 under ·3 29

```
37·7*
   2
----
75·4
39·3
----
```

$$\frac{2b+a}{10} = 11{\cdot}47$$

Opposite 11 under 7................ 63

```
     3531
       29
       63
     ----
     3623      r = 3/2
        3
     -----
 2)  10869
     ------
     5434·5
     3105·66
     -------
```

Content 8540·16 cubic yards.

EXAMPLE UNDER THE GENERAL FORMULA.

Let $a = 7{\cdot}1$; $b = 6{\cdot}8$; $c = 5{\cdot}3$; $d = 7{\cdot}6$; $e = 11{\cdot}5$; these depths are at 100 ft. apart; the breadth of the roadway = 30 ft; and the side slopes 3 : 2, or $\frac{3}{2}$: 1.

$$r = \tfrac{3}{2};\ m = 30;\ n = 100.$$

General formula $\frac{mn}{2}\left\{ a + e + 2\,(b + c + d) \right\} + \frac{nr}{3}\left\{ (a + b)^2 + (b + c)^2 + (c + d)^2 + (d + e)^2 - (ab + bc\ cd + de) \right\}$.

```
    58
   100 = n
  5800
    30 = m
2) 174000
   87000
```

```
b =  6·8
c =  5·3
d =  7·6
    19·7
       2
    39·4
     7·1 = a
    11·5 = e
    58·0
```

$$\begin{aligned}
(a + b)^2 &= (7{\cdot}1 + 6{\cdot}8)^2 = 13{\cdot}9^2 = 193{\cdot}21\\
(b + c)^2 &= (6{\cdot}8 + 5{\cdot}3)^2 = 12{\cdot}1^2 = 146{\cdot}41\\
(c + d)^2 &= (5{\cdot}3 + 7{\cdot}6)^2 = 12{\cdot}9^2 = 166{\cdot}41\\
(d + e)^2 &= (7{\cdot}6 + 11{\cdot}5)^2 = 18{\cdot}1^2 = 327{\cdot}61\\
& \qquad\qquad\qquad\qquad\qquad\quad 833{\cdot}64
\end{aligned}$$

$$\begin{aligned}
48{\cdot}28 &= a \times b = 7{\cdot}1 \times 6{\cdot}8\\
35{\cdot}04 &= b \times c = 6{\cdot}8 \times 5{\cdot}3\\
40{\cdot}28 &= c \times d = 5{\cdot}3 \times 7{\cdot}6\\
87{\cdot}40 &= d \times e = 7{\cdot}6 \times 11{\cdot}5\\
211{\cdot}00 &
\end{aligned}$$

```
      833·64
      211·00
      622·64
         100
     62264·
         3/2 = r
2)   186792
3)    93396
      31132
Add   87000
     118132 cubic feet,
```

The formula is convenient when the numbers are small and a table of the squares and products of numbers convenient.

PROBLEM.

To find the solid content of a Railroad cutting or embankment, when great accuracy is required.

RULE.

Add together the area of two parallel cross sections, and 4 times the area of a section half way between and parallel to them; and multiply the sum by one-sixth of the length measured perpendicularly to the parallel sections, and the product is the solid content required.

EXAMPLE.

Let the area of the cross section A, Fig. 71, = 2675 square feet; B = 2540 square feet; C = 2489 square feet; the distance DE = EF = 50 feet or the distance between A and B = 100; what is the content?

$$
\begin{array}{r l}
2540 & = B \\
4 & \\
\hline
10160 & \\
2675 & = A \\
2489 & = C \\
\hline
6)15324 & \\
\hline
2554\cdot & \\
100\cdot & \\
\hline
255400\cdot & \text{cubic feet.}
\end{array}
$$

$$\frac{255400}{27} = 9459{\cdot}2 \text{ cubic yards.}$$

The tables will determine the content with the same accuracy as the general rule just given, without the middle area being given.

EXAMPLE.

Let the areas of the two ends of a cutting be 4990 and 1294 square feet, the bottom width 30 feet, the length 1·60 chains, and the ratio of the slopes 1½ to 1; required the content of the cutting in cubic yards by referring to the tables.

In applying the tables to such examples, the square roots of the areas, to where the slopes meet, must be first extracted, or, which is more easy, taken from a table of square roots.

Fig. 71.

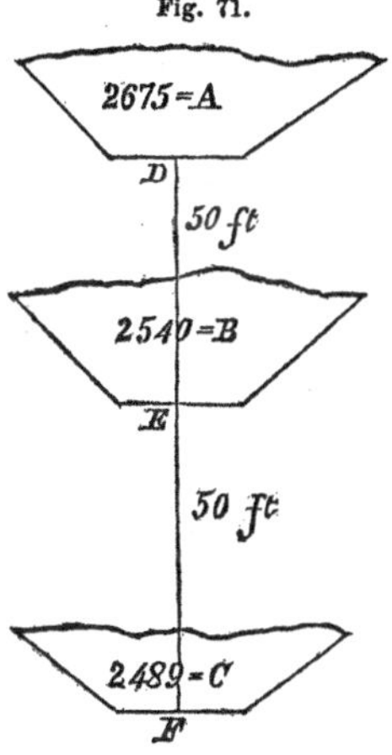

$$1\tfrac{1}{2} : 1 :: \frac{30}{2} : 10 \text{ feet,}$$

the depth below the roadway where the slopes meet,

$$\frac{30 \times 10}{2} = 150 \text{ square feet,}$$

the area of the triangle, to be added to the areas of the cross sections.

4990		1294	
150		150	
5140	and	1444	the

areas of the sections to where the slopes meet. The square roots of these numbers are 71·7 and 38· respectively.

By Table III., for 71 and 38, ... 7483

By Table IV., for $\frac{71 \times 2 + 38}{10}$ = 18· and (·7),.... 103

Content to the intersection of slopes, ... 7586

Take the content from roadway to where the slopes meet =

$\frac{150 \times 66}{27}$ =.................... $366\tfrac{2}{3}$

Content for one chain, ... $7219\tfrac{1}{3}$

$7219\tfrac{1}{3} \times 1{\cdot}60 = 11550{\cdot}933$ cub. yds.

EXAMPLE.

Let the areas of the two ends of a cutting be 3645 and 4036 square feet; the bottom width 27·2 feet; the length 17·6 feet; the ratio of the slopes 1·7 to 1; required the content of this cutting by the help of the tables.

$$1·7 : 1 :: \frac{17·2}{2} : 8 \text{ feet},$$

the depth below the roadway where the slopes meet.

$$\frac{27·2 \times 8}{2} = 108·8 \text{ square feet},$$

to be added to the areas of the cross sections. The solid content of the wedge below the roadway to where the slopes meet, for a chain = 66, in length =

$$\frac{108·8 \times 66}{27} = 265·9 \text{ cubic yards}.$$

It is necessary to make these little preliminary calculations before applying the tables, in such general examples as the one I have given above.

$$\begin{array}{rr} 3645 & 4036 \\ 108·8 & 108·8 \\ \hline \end{array}$$

$$\sqrt{3753·8} = 61·3; \quad \sqrt{4144·8} = 64·4.$$

When one or both the given depths, or square roots, exceed the limits of Table III., find the content corresponding to half the two depths and four times the result will be the content required.

By Table III, for 64 and 61, ... 9550·

Table IV., $\frac{61·3 \times 2 + 64·4}{10}$ = 19· and ·3, ... 47·

$\frac{64·4 \times 2 + 61·3}{10}$ = 19 and ·4, ... 62·

Content to intersection of slopes, ... 9659·
Content from roadway to the intersection of slopes, 265·9

9393·1

Cubic yds.
9393·1 for 66 feet
4696·5

14089·6
140·9
1·4

142·319 for 100 feet
142·319 for 1 foot.

$$142·319 \times 17·6 = 2504·8140,$$

the content in cubic yards for the distance given in the example.

To find the content of a railroad cutting, when the slopes of the two sides are different. *Rule* :—Find the content as if one of the slopes were given; take half the result and add it to half the content found by supposing the other slope only given; the sum will be the content required.

EXAMPLE.

One side of a cutting has a slope of 1½ : 1, the other side a slope of 1¼ : 1; the heights of the equivalent level cross sectional areas, taken 100 feet apart, are 13 and 11 feet; the breadth of the roadway = 30 feet; what is the content in cubic yards?

```
   11
   13
 2)24
   12
   30
  360.

For 300·, ................ 1111·111
     60·, ................  222·222
                           1333·333 cubic yards.
```

in the central part.

For 13 and 11 in Table III., 353· for 66 feet.

```
  1¼             353
  1½              11
2)2¾          8)3883
  1⅜ = 11/8      485.4
                   4·85
                   4
Slope 1⅜ : 1, length 100 feet,   490·3 cubic yards.

         1333·3
          490·3
Content  1823·6 cubic yards.
```

INVESTIGATION

OF THE

GENERAL FORMULA FOR CALCULATING THE CONTENT OF CUTTINGS AND EMBANKMENTS.

Let ABCDEFGHIJKL be a railroad cutting; the planes ABCF and LGIJ perpendicular to the plane of the roadway IJBA. Then DB = EA = b and KJ = IH = a are perpendicular to IA and BJ.

The sides JLCB and IGFA slope till their bases are to their perpendiculars as $r : 1$. AE and BD are perpendicular to CF, and JK and IH perpendicular to GL.

$$\text{BD} : \text{DC} :: 1 : r,$$

the same proportion holds in the other right-angled triangles AEF, IGH, and KJL.

$$\therefore \text{EF} = \text{DC} = br.$$
$$\text{GH} = \text{KL} = ar.$$

The length of the cutting JB = IA = n, the breadth of the roadway AB = IJ = m. Fig. 72.

Fig. 72.

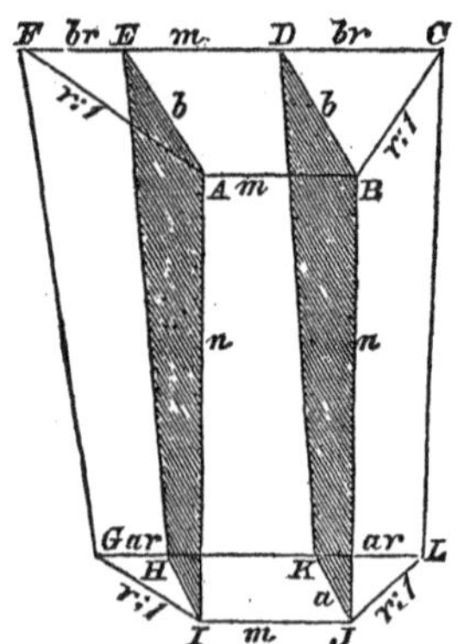

Fig. 73.

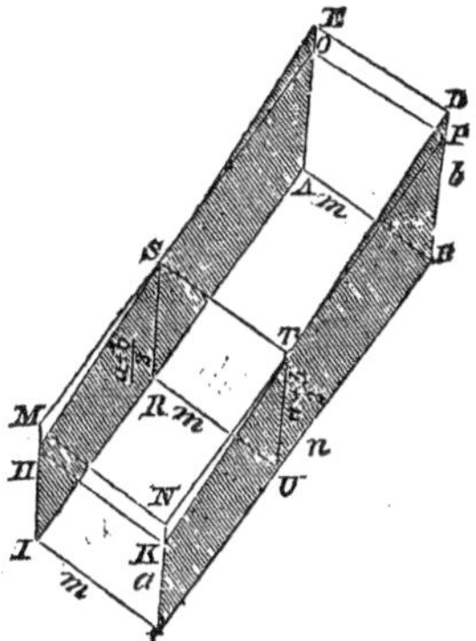

The two slopes of Fig. 73 put together make up the frustum of a pyramid FBCLJG represented in Fig. 74. The centre part EDBAIF is given in Fig. 73, the content of which I will find first.

Fig. 74.

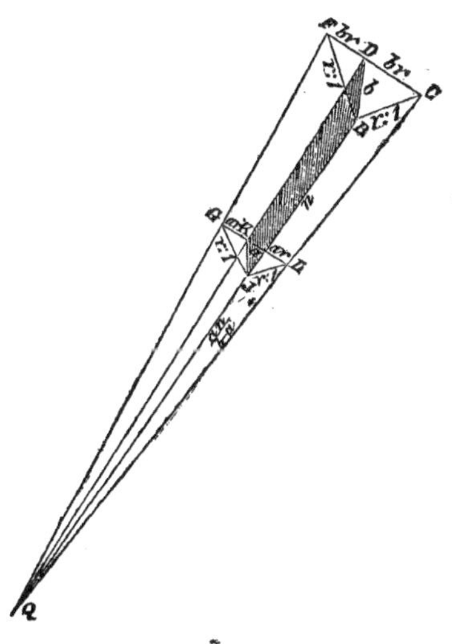

Let the plane RSTU, Fig. 73, be parallel to the ends and half-way between them; and MSOPTN parallel to the road plane ABIJ; then the solid IHEDKJ = the solid IMOPKJ =

$$\frac{a+b}{2} \times m \times n = \frac{mn}{2}(a+b).$$

From this formula Tables I. and II. are calculated, taking $n = 66$ and $n = 100$, and dividing by 27 to reduce the content to cubic yards. I will next find the content of the two slopes, Fig. 74.

$$b - a : n :: a : \frac{na}{b-a} = \text{JQ}.$$

$$\text{BQ} = n + \frac{na}{b-a} = \frac{nb}{b-a}.$$

b^2r = area of the triangle BCF.

$$\tfrac{1}{3} \times \frac{nb}{b-a} \times b^2r = \text{content of the pyramid FBCQ} =$$

$$\frac{nr}{3}\left(\frac{b^3}{b-a}\right).$$

Area of the triangle JGL $= a^2r$, hence the content of the pyramid QGLJ =

$$\tfrac{1}{3} \times \frac{na}{b-a} \times a^2r = \frac{nr}{3}\left(\frac{a^3}{b-a}\right).$$

Consequently, the solid content of the frustum JGLCFB =

$$\frac{nr}{3}\left(\frac{b^3-a^3}{b-a}\right) = \frac{nr}{3}(b^2+ab+a^2)$$

$$= \frac{nr}{3}\left\{(a+b)^2 - ab\right\}.$$

From this expression, Table III. has been calculated, taking $n = 66$ feet and dividing by 27 to reduce the content to cubic yards. The general formula is readily found by taking the sum of the expressions.

$$\frac{mn}{2}(a+b) + \frac{nr}{3}\left\{(a+b)^2 - ab\right\}$$

$$\frac{mn}{2}(b+c) + \frac{nr}{3}\left\{(b+c)^2 - bc\right\}$$

$$\frac{mn}{2}(c+d) + \frac{nr}{3}\left\{(c+d)^2 - cd\right\}$$

&c. + &c.,

which becomes,

$$\frac{mn}{2}\left\{a + z + 2(b + c + d \ldots) + \right.$$

$$\frac{nr}{3}\left\{(a + b)^2 + (b + c)^2 \times (c + d)^2 + \ldots\right.$$

$$\left. - [ab + bc + cd + \ldots]\right\},$$

the general formula that I proposed to demonstrate. I will add an example that often occurs in practice; when cuttings or embankments are measured after the work is done, the sides have different slopes from one another, and from those intended to be given, to find the content of such, the following rule may be useful.

RULE.

Find the content of the centre portion, as in the preceding examples; in finding the content of the two slopes, employ half the sum of the ratios (the consequents being unity), instead the constant ratio used in other cases.

EXAMPLE.

Let the bottom width = 36 feet; the depths of the level equalized cross sections = 20 and 30 feet respectively; one of the side-slopes $1\frac{1}{2}$ to 1, the other 2 to 1; what is the content, in cubic yards, for a length of 100 feet?

$$\begin{array}{r} 2 \;:\; 1 \\ 1\frac{1}{2} \;:\; 1 \\ \hline 2)3\frac{1}{2} \quad\quad \\ \hline 1\frac{3}{4} \;:\; 1 = r : 1. \end{array}$$

BY THE FORMULA.

$$\frac{36 \times 100}{2} \times (20 + 30) = 90000,$$ cubic feet in the central part.

$$\begin{array}{r} 20 \\ 30 \\ \hline 50^2 = 2500 = (a + b)^2 \\ 20 \times 30 = 600 \\ \hline 1900 = ab + a^2 + b^2 \\ 100 \\ \hline 190000 \\ 7 \quad r = \frac{7}{4} \\ \hline 4)\ 1330000 \\ \hline 3)\ 332500 \\ \hline 110833.3 \end{array}$$

110833.3 cubic feet in the side slopes.

```
       110833·3
        90000·
     ----------
27)  200833·3  (7438·2
     189
     ---
      118
      108
      ---
       103
        81
       ---
        223
        216
        ---
          73
          54
          --
```

BY THE TABLES.

Table III.

For 30 and 20 will be found 1548; this is for a length of 66 feet.

```
                      1548
                       774
                      ----
for 99 feet  —        2322
                        23 22
                           23
                      -------
for 100 ft.           2345·45
                            7      r = 7/4
                      --------
               4)    16418·15
                      --------
                      4104·54
                      3333·33 found in Table II. for 9000.
                      -------
Cubic yards           7437·87
                      =======

          20
          30
          --
      2)  50
          --
          25 × 36 = 9000.
```

By inspecting the tables the content is found to be 7437·87 cubic yards; by the formula the content is 7438·2 cubic yards, the difference is less than half a cubic yard.

ELEVATION OF THE EXTERIOR RAIL

IN CURVES TO COUNTERACT THE CENTRIFUGAL FORCE CAUSED BY THE VELOCITY OF THE TRAIN.

Let w be the weight of the moving body or train, v its velocity in feet a second, R the radius of the curve, and g = force of gravity at the surface of the earth. Representing the centrifugal force by f, by a well-known dynamical expression,

$$f = \frac{wv^2}{g\text{R}}$$

If R = one quarter of a mile = 1320 feet, v = velocity = 38·6 feet a second, and $g = 32\frac{1}{6}$ feet,

$$\text{Then } f = \frac{w \times (38{\cdot}6)^2}{32\frac{1}{6} \times 1320} = \frac{193}{5500} w;$$

In this example the centrifugal force is $\frac{1}{28}$ part of the weight.

If R = a mile = 5280 feet, v = 60 miles an hour = 88 feet a second, g as before $32\frac{1}{6}$ feet,

$$f = \frac{w \times 88^2}{32\frac{1}{6} \times 5280} = \frac{1}{22} w;$$

In this last case, the centrifugal force that urges the moving body to leave the curve $\frac{1}{12}$ of its weight.

This force is in a great measure counteracted by the conical tread of the wheels, each pair of which is firmly fixed on an axle, as AB, Fig. 75.

Fig. 75.

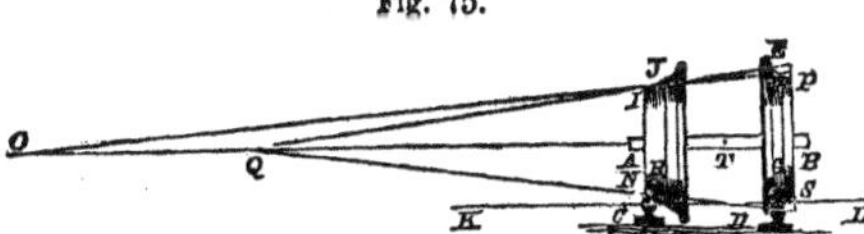

The rails C and D may be level, yet if the points of contact be at I and E, coned wheels will run round a circle whose centre is at Q. If the points of contact of the rails be J and F, then the centre will be at O; but if the wheels assume the position that the equal circles G and H come in contact with the rails C and D, the axle AB and KL would be parallel, and hence do not meet.

The coned tread, the lateral play of the flanges, about half an inch on each side, and the centrifugal force of the weight moved on the curve, enlarge the diameter of the exterior wheel B, and diminish that of the interior A; hence there is a centripetal force directed towards the centres of the cones Q, O, &c., as the rails change their points of contact on the coned tires.

Let d be the diameter of the wheels, at the circles G and H, when they stand level, or rather when the line K L joining the points of contact is parallel to the axis AB.

Let the outer diameter be increased by a variable small quantity z, as the rail touches, nearer the flange; or let the outer diameter become

$$d + z.$$

It is clear that the inner diameter will become

$$d - z.$$

The value of z, generally varies from 0 to $\frac{3}{5}$ of an inch; so that the diameters of the wheels may be made to vary according to the circumstances of the motion from about 36 to 36·6 inches, sometimes more and sometimes less.

Let r be the variable radius OT, answering to the increase z, and b = gage or breadth of the road CD; then $r + \frac{1}{2}b$ and $r - \frac{1}{2}b$ are the distances of the circles of contact from the centre O. By similar triangles

$$d + z : d - z :: r + \frac{b}{2} : r - \frac{b}{2}$$

$$\therefore d : z :: 2r : b$$

$$r = \frac{bd}{2z}.$$

If the breadth of the road = 4·7 feet the diameter of the wheel, = 3 feet, and $z = \frac{1}{10}$ of an inch.

$$\text{OT} = r = \frac{12 \times 4{\cdot}7 \times 3 \times 12}{{\cdot}2} = 10152 \text{ inches} = 846 \text{ feet.}$$

Again, for the sake of example, let $z = \frac{1}{40}$ of an inch, the breadth of the road = 7 feet = 84 inches, the diameter of the wheel at a level tread = 30 inches; required OT = r.

$$r = \frac{30 \times 84}{{\cdot}05} = 50400 \text{ inches} = 4200 \text{ feet.}$$

But the centripetal force corresponding to radius r, is

$$p = \frac{wv^2}{gr},$$

which acts in a contrary direction to the centrifugal force, they will hold each other in equilibrium when they become equal, and the train will have no inducement to fly off the track.

$$\therefore \frac{wv^2}{gr} = \frac{wv^2}{gR},$$

$$\text{Or } r = R.$$

Consequently the vertex O, of the imaginary cone must coincide with the centre of the curve of the railroad to avoid slipping or dragging. But it was before shown that

$$r = \frac{bd}{2z}$$

$$\therefore R = \frac{bd}{2z},$$

$$z = \frac{bd}{2R},$$

which is the increment and decrement that the exterior and interior wheels respectively receive to produce the equilibrium between the centripetal and centrifugal forces of the train.

Let the gage of the road $= b = 64$ inches, the diameter of the wheels at the points where they rest level on the rails $= 36$ inches, R, the radius $= 600$ feet, $2R \times 12 = 14400$ inches.

Hence in this case

$$z = \frac{64 \times 36}{14400} = \cdot 16 \text{ of an inch.}$$

The coning of the wheels will not compensate for much more than this as there is only a play of about half an inch on each side between the flanges and the rails.

Let Fig. 77 represent part of the tire of a railway car wheel; the rise from A to C half an inch in $3\frac{1}{2}$ inches, that is,

$$AB : BC : 7 : 1.$$

But in order that the statement may be general and applicable to all cases, let

$$AB : BC :: n\ 1.$$

When the wheels stand level on level rails, let E be one of the points of contact, DB the space given for the play of the wheel between the flange and the rail; this, for the sake of example I put $= \frac{1}{2}$ an inch $= v$.

The rail and wheel may touch at any other point between C and A, yet the space DB is unaltered, and the elevation will be as n to 1. The diameter through E is put $= d$.

Fig. 76.

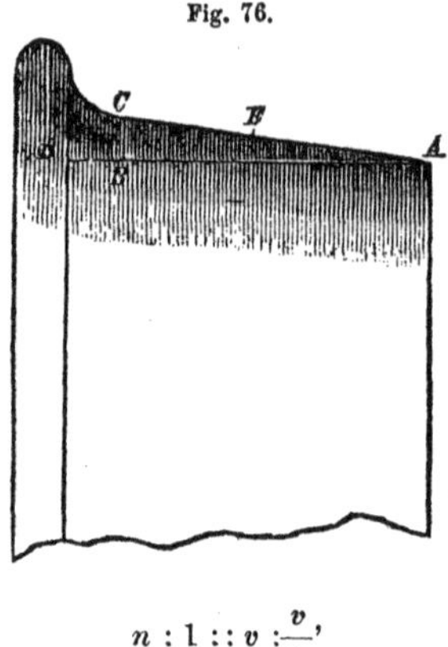

$$n : 1 :: v : \frac{v}{n},$$

$$\therefore \frac{2v}{n} = z$$

Giving to v what I have just supposed to be its greatest value $= \frac{1}{2}$ an inch.

$$z = \tfrac{1}{7},$$

$$\text{and } r = \frac{ndb}{4v} = \frac{bd}{2z}.$$

It is clear as z increases r decreases.

$$\text{Let } d = 3 \text{ feet, } b = 4{\cdot}7 \text{ feet,}$$

$$z = \tfrac{1}{7} \text{ of an inch} = \tfrac{1}{84} \text{ part of a foot.}$$

$$r = 3 \times 4{\cdot}7 \times \frac{84}{2} = 592{\cdot}2 \text{ feet}$$

the least possible radius of curvature on the suppositions made, in which the two forces balance each other, supposing the two rails to be exactly level. I will assume another case,

$$\text{Let } v = \tfrac{1}{4} \text{ an inch, and } \frac{1}{n} = \tfrac{1}{7};$$

$$z = \frac{2v}{n} = \tfrac{1}{14} \text{ of an inch.}$$

$$r = \frac{bd}{2z} = 3 \times 4{\cdot}7 \times \frac{14}{2} \times 12 = 1184{\cdot}4 \text{ feet.}$$

Again let $v = \frac{1}{8}$ of an inch, $z = \frac{2v}{n} = \frac{1}{28}$

$$2z = \frac{1}{168} \text{ of a foot,}$$

$$r = 3 \times 4{\cdot}7 \times \frac{168}{11} = 2368{\cdot}8 \text{ ft.}$$

Hence it may be inferred that the coning of the wheels, without a rise of the outer rail, cannot be depended on in curves of less than 2000 feet radius.

What I have said on this subject up to this is mere preliminary matter, I will now find the elevation of the exterior rail for any radius R of a railroad curve.

Let $x =$ the required elevation, b, as before, the gage of the road,

$$\frac{wx}{b},$$

will be the force drawing the train to the interior rail on account of the elevation x. This force must hold the centrifugal force in equilibrium, hence

$$\frac{wx}{b} = \frac{wv^2}{gR}$$

$$\therefore x = \frac{bv^2}{gR}.$$

EXAMPLE.

Let $R = 1910$ feet, $g = 32{\cdot}2$, $b = 4{\cdot}7$, $v = 44$ feet a second $=$ 30 miles an hour.

$$x = \frac{4{\cdot}7 \times 44^2}{23{\cdot}2 \times 1910} = {\cdot}224$$

This value of x involves the elevation given by the *coning* of the wheels; the proper coning for a radius of 1,910 ft., or

$$z = \frac{bd}{2R},$$

as before established,

$$= \frac{36 \times 56\frac{1}{2}}{2 \times 12 \times 1910} = {\cdot}044.$$

$b = 36$, $d = 2$ ft. $8\frac{1}{2}$ in.

From ·224
Take ·044

·180 inches.

When the coning of the wheels is so defective that the proper circles on the wheels cannot be employed, then the outer rail is pressed and broken by the flange. It is in this particular that engineers have erred. Wheel-makers have no system of coning so that the wheels they make will run on a curve with a given velocity without injuring the flange and outer rail. I here, for the first time, lay down the means of doing so; it is simple, when the rise (x) of the outer rail is formed for any radius (R), and velocity (v); if the coning (z) cannot be established more or less without constraint, the wheels will not run on the curve with freedom. In fact, at all speeds on every curve, the apices O, Q, &c., Fig. 76, must coincide with the centre of the railroad curve, or damage will be done to the running gear. Want of skill in this particular is made up by a range given to the king-bolt, a lateral-motion beam, and other contrivances.

EXAMPLE.

Let the diameter of the wheels $= d = 33\frac{1}{2}$ inches, at the points where they touch on level rails; say the coning allows the diameter of the outer wheels to be increased, while the diameter of the inner wheels are diminished ·13 inches; what is the least radius, with this play, that these wheels will accommodate themselves to; and what is the elevation of the outer rail for a velocity of 50 miles an hour on a track of 819 feet radius, the guage = 4 feet $8\frac{1}{2}$ inches?

$$R = \frac{bd}{2z}.$$

$$\therefore R = \frac{33{\cdot}5 \times 56{\cdot}5}{2 \times .13} = 7{,}280 \text{ inches} = 606 \text{ feet } 8 \text{ inches},$$

the least radius that these wheels will accurately accommodate.

50 miles = 264,000 feet,
$73\frac{1}{3}$ feet a second.

$$x = \frac{bv^2}{gR}$$

$$= x = \frac{4{\cdot}7 \times (73\frac{1}{3})^2}{32{\cdot}2 \times 819} = \frac{4{\cdot}7 \times 220^2}{9 \times 32{\cdot}2 \times 819} = {\cdot}958.$$

For a radius of 819 feet,

$$z = \frac{bd}{2R} = \frac{2{\cdot}8 \times 4{\cdot}7}{2 \times 819} = {\cdot}008.$$

·958 — ·008 = ·95 elevation of the outer rail.

LEVELLING.

LET the levelling instrument be set up and adjusted as at Fig. 77. By a plummet P, the point A, from which the observation is said to be made, is easily formed; the height of the instrument above this point A, to the centre of the telescope D, is also easily taken, when required.

Fig. 77.

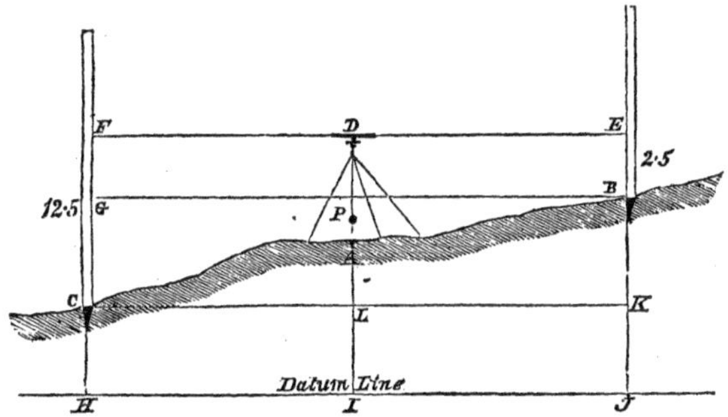

Suppose C and B to be two stations, pegs are driven even with the surface to support the levelling-staff. The positions of these pegs are generally marked by stakes two feet in length, driven near them and numbered.

Turn the instrument to the staff placed at B, and say the hairs in the telescope cut 2·5 — BE; when the staff is placed at C, let CF = 12·5. Then it is clear that

$$12{\cdot}5 - 2{\cdot}5 = 10 \text{ feet} = CG,$$

the height of the point B above C. If DA, the height of the instrument = 5 feet, 12·5 — 5 = 7·5 feet, the height of the point A above C, and 5· — 2·5 = 2·5 = height of the point B above A. The line of sight, EF, termed the *line of collimation,* is supposed to be horizontal in all directions that may be given to the telescope when the instrument is properly made and adjusted.

Mathematical instrument-makers resort to all sorts of tricks to make their levels appear correct; on most of them, but little reliance can be placed. However, when the level is stationed at A, half way between B and C, all errors are balanced, whether instrumental or otherwise. It is clear if C F be made to appear a foot higher or lower, BE will be affected in the same manner by the same error in the line of *collimation.*

If HJ be the datum line, and suppose the station C to be 7 feet above it.

FH = 19·5 feet = DJ.
14·5 = 19·5 — 5 = AJ.
17· = 19·5 — 2·5 = JB.

Let the datum-line, Fig. 78, be five feet below station 214; by the way of example, set the instrument at B, and adjust it; stand the levelling-staff on the peg at station 214; let the line of collimation AC strike the staff four feet from the peg, so that the point

Fig. 78.

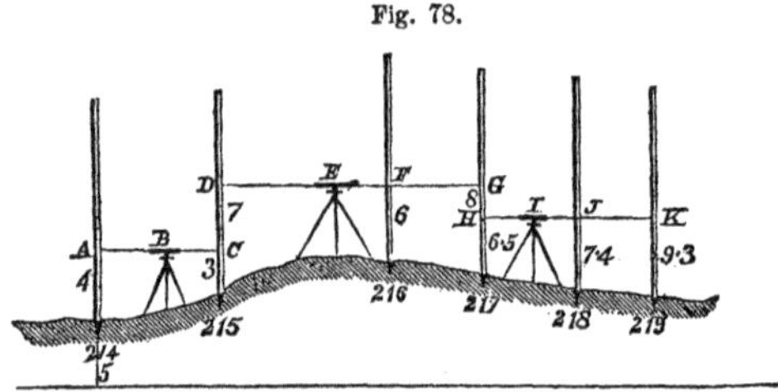

Datum-line, high tide, Battery, New York City.

A and the centre of the telescope B are 9 feet from the datum-line. By placing the staff on the peg of station 215, and turning the telescope, the point C is found to be 3 feet above 215, hence station 215 is 1 foot above station 214.

$$5 + 4 - 3 = 6,$$

the distance of the surface of the ground, at station 215, from the datum-line. By continuing the process called levelling, and by placing the staff at stations 216, 217, 218, 219, and the instrument at E and I, the heights of the stations and of the instrument, that is of the line of collimation, above the datum-line, are worked out in lengthy detail thus:

$$6 + 7 = 13 \text{ feet},$$

the distance of D, or of the telescope E from the datum-line or of the points F and G.

$$6 + 7 - 6 = 7 \text{ feet},$$

the distance of station 216 from the datum-line

$$6 + 7 - 8 = 5 \text{ feet},$$

the distance of station 217 from the datum.

$$5 + 6{\cdot}5 = 11{\cdot}5 \text{ feet},$$

the height of H, the instrument I, and the points J and K from the datum.

$$11{\cdot}5 - 7{\cdot}4 = 4{\cdot}1 \text{ feet},$$

the height of station 218 above the datum.

$$11{\cdot}5 - 9{\cdot}3 = 2{\cdot}2 \text{ feet},$$

height of station 219 above datum.

Consequently, as a general rule, add for each new position of the instrument the back reading of the staff, to the found elevation of the point where the staff stands, for the height of the telescope of the instrument; and subtract from each height thus found the readings of the staff placed at new stations in advance to find the heights of those stations above the datum-line. This work, put in field-book form, will stand as follows:

Stations.	Recorded from the Staff.	Height of Telescope Collimation.	Height above the Datum.
214	4·00	9·00	+ 5·00
215	3·00		+ 6·00
	7·00	13·00	
216	6·00		+ 7·00
217	8·00		+ 5·00
	6·50	11·50	
218	7·40		+ 4·10
219	9·30		+ 2·20

In levelling cross-sections and staking side slopes the succeeding form of level-field-book I have found very convenient, but each engineer has a method of his own. The level cutting, DEHF, Fig. 79, is not one of equal cross-sectional area to ACBHF, but merely a cutting formed by level-line passing over the peg C. The level-line that reduces the cross section to one of equal area

Fig. 79.

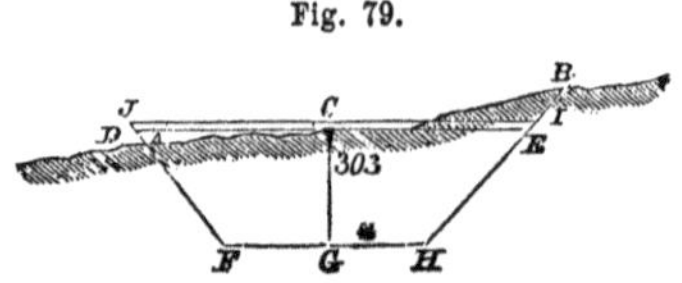

is in this case higher than DE; the level cross section JIHF = in area the irregular cross section ACBHF.

	Number of Station, As C, Fig. 79.	Depth of Cuttings or Embankments. CG, Fig. 79.	Computed half width CD = CE. Fig. 79.	Corrected half-widths for edge of cutting or foot of Embankment. Left, or AC.	Right, or CB.
		FEET.	FEET.		
CUTTING.	303	30·00	60·00	52·60	71·70
	304	3·00	19·50	19·44	36·11
	305	28·00	57·00	57·95	56·08
	306	19·68	44·42	44·42	44·42
	307	20·00	45·00	32·60	83·33
EMBANKMENT.	308	16·08	39·12	39·72	59·12
	309	30·00	60·00	54·24	68·05
	310	32·18	63·27	63·00	63·62

It may happen that the line JI bounding the equivalent level-cutting may coincide with DE, or be lower than it. I am particular on this point, as no writer on this subject mentions the circumstance.

MISCELLANEOUS PROBLEMS

OF THE BENDING OF RAILS TO SUIT DIFFERENT CURVES.

When the length of an arc $= x$, to radius 1, the cosine of such arc is well known to be

$$1 - \frac{x^2}{1 \cdot 2} + \frac{x^4}{1 \cdot 2 \cdot 3 \cdot 4} - \&c.$$

The versine of the same arc =

$$\frac{x^2}{1 \cdot 2} - \frac{x^4}{1 \cdot 2 \cdot 3 \cdot 4} + \&c.$$

Let the length of the rail before being bent $= 2n$, and the radius of the curve to which it has to be bent $= r$.

$$\frac{n}{r} = \text{length of half the arc reduced to radius 1.}$$

Putting this for x, the versine becomes

$$\frac{n^2}{2 \cdot r^2} - \frac{n^4}{24 \cdot r^4} + \&c.$$

But because $\frac{n}{r}$ is always very small to reject $\frac{n^4}{24 \cdot r^4}$ would not affect the result in the fifth decimal; hence the middle ordinate for curving a rail whose length $= 2n$ to a circle whose radius $= r$, will be $\frac{n^2}{2r2}$ to radius 1, or

$$\frac{n^2}{2r} \text{ to radius } r.$$

If the chord = 100 feet, and the deflection-angle $= \theta$, then $r = \frac{50}{\sin. \theta}$, and the middle ordinate of the bent rail becomes

$$n_2 \times \frac{\sin. \theta}{100}.$$

EXAMPLE.

Let $r = 800$ feet, $2n$, the length of the rail $= 20$ feet. What is the ordinate in the middle of this rail bent to suit this curve.

$$n^2 = 10^2 = 100;$$
$$2r = 1600;$$
$$\frac{n^2}{2r} = \frac{100}{1600} = \cdot 0625 \text{ feet} = \cdot 75 \text{ inches} = \tfrac{3}{4} \text{ of an inch.}$$

EXAMPLE.

If the chord $= 100$ feet, and its deflection from a tangent at one of its extremities $= 3° \ 35'$, what is the ordinate in the middle of a rail 20 feet long when bent

$$n = \frac{20}{2} = 10.$$
$$\text{Natural sine } 3° \ 35' = \cdot 0625.$$
$$\therefore n^2 \times \frac{\sin. \theta}{100} = \cdot 0625 \text{ feet,}$$

the ordinate when the rail is bent.

EXAMPLE.

What is the ordinate in the middle of a rail 24 feet long, bent to suit a 4° curve, that is a curve, or circular arc, in which a chord of 100 feet subtends 4° at the centre of the circle.

$$n^2 = 12^2 = 144.$$
$$\text{Nat. sin. } 2° = \cdot 0349.$$

Middle ordinate $= 144 \times \cdot 000349 = \cdot 050256$ ft. or about $\frac{6}{10}$ of an inch.

THE END.

INDEX.

Practical and Scientific Books,

PUBLISHED BY

HENRY CAREY BAIRD,

INDUSTRIAL PUBLISHER,

No. 406 Walnut Street,
PHILADELPHIA.

☞ **Any of the following Books will be sent by mail, free of postage, at the publication price. Catalogues furnished on application.**

American Miller and Millwright's Assistant:

A new and thoroughly revised Edition, with additional Engravings. By William Carter Hughes. In one volume, 12 mo., ...$1.25

Armengaud, Amoroux, and Johnson.

THE PRACTICAL DRAUGHTSMAN'S BOOK OF INDUSTRIAL DESIGN, and Machinist's and Engineer's Drawing Companion; forming a complete course of Mechanical Engineering and Architectural Drawing. From the French of M. Armengaud the elder, Prof. of Design in the Conservatoire of Arts and Industry, Paris, and MM. Armengaud the younger, and Amouroux, Civil Engineers. Rewritten and arranged, with additional matter and plates, selections from and examples of the most useful and generally employed mechanism of the day. By William Johnson, Assoc. Inst. C. E., Editor of "The Practical Mechanic's Journal." Illustrated by fifty folio steel plates and fifty wood-cuts. A new edition, 4to.,....$9.00

Among the contents are:—*Linear Drawing, Definitions and Problems*, Plate I. Applications, Designs for inlaid Pavements, Ceilings and Balconies, Plate II. Sweeps, Sections and Mouldings, Plate III. Elementary Gothic Forms and Rosettes, Plate IV. Ovals, Ellipses,

Parabolas and Volutes, Plate V. Rules and Practical Data. *Study of Projections*, Elementary Principles, Plate VI. Of Prisms and other Solids, Plate VII. Rules and Practical Data. *On Coloring Sections, with Applications*—Conventional Colors, Composition or Mixture of Colors, Plate X. *Continuation of the Study of Projections*—Use of sections—details of machinery, Plate XI. Simple applications—spindles, shafts, couplings, wooden patterns, Plate XII. Method of constructing a wooden model or pattern of a coupling, Elementary applications—rails and chairs for railways, Plate XIII. *Rules and Practical Data*—Strength of material, Resistance to compression or crushing force, Tensional Resistance, Resistance to flexure, Resistance to torsion, Friction of surfaces in contact.

THE INTERSECTION AND DEVELOPMENT OF SURFACES, WITH APPLICATIONS.—*The Intersection of Cylinders and Cones*, Plate XIV. *The Delineation and Development of Helices, Screws and Serpentines*, Plate XV. Application of the helix—the construction of a staircase, Plate XVI. The Intersection of surfaces—applications to stop-cocks, Plate XVII. *Rules and Practical Data*—Steam, Unity of heat, Heating surface, Calculation of the dimensions of boilers, Dimensions of firegrates, Chimneys, Safety-valves.

THE STUDY AND CONSTRUCTION OF TOOTHED GEAR.—Involute, cycloid, and epicycloid, Plates XVIII. and XIX. Involute, Fig. 1, Plate XVIII. Cycloid, Fig. 2, Plate XVIII. External epicycloid, described by a circle rolling about a fixed circle inside it, Fig. 3, Plate XIX. Internal epicycloid, Fig. 2, Plate XIX. Delineation of a rack and pinion in gear, Fig. 4, Plate XVIII. Gearing of a worm with a worm-wheel, Figs. 5 and 6, Plate XVIII. *Cylindrical or Spur Gearing*, Plate XIX. Practical delineation of a couple of Spur-wheels, Plate XX. *The Delineation and Construction of Wooden Patterns for Toothed Wheels*, Plate XXI. *Rules and Practical Data*—Toothed gearing, Angular and circumferential velocity of wheels, Dimensions of gearing, Thickness of the teeth, Pitch of the teeth, Dimensions of the web, Number and dimensions of the arms, wooden patterns.

CONTINUATION OF THE STUDY OF TOOTHED GEAR.—Design for a pair of bevel-wheels in gear, Plate XXII. Construction of wooden patterns for a pair of bevel-wheels, Plate XXIII. *Involute and Helical Teeth*, Plate XXIV. *Contrivances for obtaining Differential Movements*—The delineation of eccentrics and cams, Plate XXV. *Rules and Practical Data*—Mechanical work of effect, The simple machines, Centre of gravity, On estimating the power of prime movers, Calculation for the brake, The fall of bodies, Momentum, Central forces.

ELEMENTARY PRINCIPLES OF SHADOWS.—*Shadows of Prisms, Pyramids and Cylinders*, Plate XXVI. *Principles of Shading*, Plate XXVII. *Continuation of the Study of Shadows*, Plate XXVIII. *Tuscan Order*, Plate XXIX. *Rules and Practical Data*—Pumps, Hydrostatic principles, Forcing pumps, Lifting and forcing pumps, The Hydrostatic press, Hydrostatical calculations and data—discharge of water through different orifices, Gaging of a water-course of uniform section and fall, Velocity of the bottom of water-courses, Calculation of the discharge of water through rectangular orifices of narrow edges, Calculation of the discharge of water through overshot outlets, To determine the width of an overshot outlet, To determine the depth of the outlet, Outlet with a spout or duct.

APPLICATION OF SHADOWS TO TOOTHED GEAR, Plate XXX. *Application of Shadows to Screws*, Plate XXXI. *Application of Shadows to a Boiler and its Furnace*, Plate XXXII. *Shading in Black—Shading in Colors*, Plate XXXIII.

THE CUTTING AND SHAPING OF MASONRY, Plate XXXIV. *Rules and Practical Data*—Hydraulic motors, Undershot water wheels, with plane floats and a circular channel, Width, Diameter, Velocity, Number and capacity of the buckets, Useful effect of the water wheel, Overshot water wheels, Water wheels with radial floats, Water wheel with curved buckets, Turbines. *Remarks on Machine Tools.*

THE STUDY OF MACHINERY AND SKETCHING.—Various applications and combinations: *The Sketching of Machinery*, Plates XXXV. and XXXVI. *Drilling Machine; Motive Machines;* Water wheels, Construction and setting up of water wheels, Delineation of water wheels, Design for a water wheel, Sketch of a water wheel; *Overshot Water Wheels. Water Pumps*, Plate XXXVII. *Steam Motors;* High-pressure expansive steam engine, Plates XXXVIII., XXXIX. and XL. *Details of Construction; Movements of the Distribution and Expansion Valves; Rules and Practical Data*—Steam engines: Low-pressure condensing engines without expansion valve, Diameter of piston, Velocities, Steam pipes and passages, Air-pump and condenser, Cold-water and feed-pumps, High-pressure expansive engines, Medium pressure condensing and expansive steam engine, Conical pendulum or centrifugal governor.

OBLIQUE PROJECTIONS.—Application of rules to the delineation of an oscillating cylinder, Plate XLI.

PARALLEL PERSPECTIVE.—Principles and applications, Plate XLII.

TRUE PERSPECTIVE.—Elementary principles, Plate XLIII. Applications—flour mill driven by belts, Plates XLIV. and XLV. Description of the mill, Representation of the mill in perspective, Notes of recent improvements in flour mills, Schiele's mill, Mullin's "ring millstone," Barnett's millstone, Hastie's arrangement for driving mills, Currie's improvements in millstones; *Rules and Practical Data*—Work performed by various machines, Flour mills, Saw mills, Veneer-sawing machines, Circular saws.

EXAMPLES OF FINISHED DRAWINGS OF MACHINERY.—Plate A, Balance water-meter; Plate B, Engineer's shaping machine; Plate C D E, Express locomotive engine; Plate F., Wood planing machine; Plate G, Washing machine for piece goods; Plate H, power loom; Plate I, Duplex steam boiler; Plate J, Direct-acting marine engines.

DRAWING INSTRUMENTS.

Barnard (Henry). National Education in Europe:

Being an Account of the Organization, Administration, Instruction, and Statistics of Public Schools of different grades in the principal States. 890 pages, 8vo., cloth, ... $3.00

Barnard (Henry). School Architecture.

New Edition, 300 cuts, cloth, ... $2.00

Beans. A Treatise on Railroad Curves and the Location of Railroads.

By E. W. Beans, C. E. 12mo. (In press.)

Bishop. A History of American Manufactures,

From 1608 to 1860; exhibiting the Origin and Growth of the Principal Mechanic Arts and Manufactures, from the Earliest Colonial Period to the Present Time; with a

Notice of the Important Inventions, Tariffs, and the Results of each Decennial Census. By J. Leander Bishop, M. D, : to which is added Notes on the Principal Manufacturing Centres and Remarkable Manufactories. By Edward Young and Edwin T. Freedley. In two vols., 8vo. Vol. 1 now ready. Price,........................$3.00

Bookbinding: A Manual of the Art of Book binding,

Containing full instructions in the different branches of Forwarding, Gilding and Finishing. Also, the Art of Marbling Book-edges and Paper. By James B. Nicholson. Illustrated. 12mo., cloth,................................$2.00

CONTENTS.—Sketch of the Progress of Bookbinding, Sheet-work, Forwarding the Edges, Marbling, Gilding the Edges, Covering, Half Binding, Blank Binding, Boarding, Cloth-work, Ornamental Art, Finishing, Taste and Design, Styles, Gilding, Illuminated Binding, Blind Tooling, Antique, Coloring, Marbling, Uniform Colors, Gold Marbling, Landscapes, etc., Inlaid Ornaments, Harmony of Colors, Pasting Down, etc., Stamp or Press-work, Restoring the Bindings of Old Books, Supplying imperfections in Old Books, Hints to Book Collectors, Technical Lessons.

Booth and Morfit. The Encyclopedia of Chemistry, Practical and Theoretical:

Embracing its application to the Arts, Metallurgy, Mineralogy, Geology, Medicine, and Pharmacy, By James C. Booth, Melter and Refiner in the United States Mint; Professor of Applied Chemistry in the Franklin Institute, etc.; assisted by Campbell Morfit, author of "Chemical Manipulations," etc. 7th Edition. Complete in one volume, royal octavo, 978 pages, with numerous wood cuts and other illustrations,................................$6.00

From the very large number of articles in this volume, it is entirely impossible to give a list of the Contents, but attention may be called to some among the more elaborate, such as Affinity, Alcoholometry, Ammonium, Analysis, Antimony, Arsenic, Blowpipes, Cyanogen, Distillation, Electricity, Ethyl, Fermentation, Iron, Lead and Water.

Brewer; (The Complete Practical.)

Or Plain, Concise, and Accurate Instructions in the Art of Brewing Beer, Ale, Porter, etc., etc., and the Process of Making all the Small Beers. By M. Lafayette Byrn, M. D. With Illustrations. 12mo......................$1.25

"Many an old brewer will find in this book valuable hints and suggestions worthy of consideration, and the novice can post himself up in his trade in all its parts."—*Artisan.*

Builder's Pocket Companion:

Containing the Elements of Building, Surveying, and Architecture; with Practical Rules and Instructions connected with the subject. By A. C. Smeaton, Civil Engineer, etc. In one volume, 12mo., ...$1.25

CONTENTS.—The Builder, Carpenter, Joiner, Mason, Plasterer, Plumber, Painter, Smith, Practical Geometry, Surveyor, Cohesive Strength of Bodies, Architect.

"It gives, in a small space, the most thorough directions to the builder, from the laying of a brick, or the felling of a tree, up to the most elaborate production of ornamental architecture. It is scientific, without being obscure and unintelligible; and every house-carpenter, master, journeyman, or apprentice, should have a copy at hand always."—*Evening Bulletin.*

Byrne. The Handbook for the Artisan, Mechanic, and Engineer,

Containing Instructions in Grinding and Sharpening of Cutting Tools, Figuration of Materials by Abrasion, Lapidary Work, Gem and Glass Engraving, Varnishing and Lackering, Abrasive Processes, etc., etc. By Oliver Byrne. Illustrated with 11 large plates and 185 cuts. 8vo., cloth,..$5.00

CONTENTS.—Grinding Cutting Tools on the Ordinary Grindstone; Sharpening Cutting Tools on the Oilstone; Setting Razors; Sharpening Cutting Tools with Artificial Grinders; Production of Plane Surfaces by Abrasion; Production of Cylindrical Surfaces by Abrasion; Production of Conical Surfaces by Abrasion; Production of Spherical Surfaces by Abrasion; Glass Cutting; Lapidary Work; Setting, Cutting, and Polishing Flat and Rounded Works; Cutting Faucets; Lapidary Apparatus for Amateurs; Gem and Glass Engraving; Seal and Gem Engraving; Cameo Cutting; Glass Engraving, Varnishing, and Lackering; General Remarks upon Abrasive Processes; Dictionary of Apparatus; Materials and Processes for Grinding and Polishing commonly employed in the Mechanical and Useful Arts.

Byrne. The Practical Metal-worker's Assistant,

For Tin-plate Workers, Braziers, Coppersmiths, Zincplate Ornamenters and Workers, Wire Workers, Whitesmiths, Blacksmiths, Bell Hangers, Jewellers, Silver and Gold Smiths, Electrotypers, and all other Workers in Alloys and Metals. Edited by Oliver Byrne. Complete in one volume, octavo,.......................................$9.00

It treats of Casting, Founding, and Forging; of Tongs and other Tools; Degrees of Heat and Management of Fires; Welding of

Heading and Swage Tools; of Punches and Anvils; of Hardening and Tempering; of Malleable Iron Castings, Case Hardening, Wrought and Cast Iron; the Management and Manipulation of Metals and Alloys, Melting and Mixing; the Management of Furnaces, Casting and Founding with Metallic Moulds, Joining and Working Sheet Metal; Peculiarities of the different Tools employed; Processes dependent on the ductility of Metals; Wire Drawing, Drawing Metal Tubes, Soldering; The use of the Blowpipe, and every other known Metal Worker's Tool.

Byrne. The Practical Model Calculator,

For the Engineer, Machinist, Manufacturer of Engine Work, Naval Architect, Miner, and Millwright. By OLIVER BYRNE, Compiler and Editor of the Dictionary of Machines, Mechanics, Engine Work and Engineering, and Author of various Mathematical and Mechanical Works. Illustrated by numerous engravings. Complete in one large volume, octavo, of nearly six hundred pages,..$4.50

The principal objects of this work are: to establish model calculations to guide practical men and students; to illustrate every practical rule and principle by numerical calculations, systematically arranged; to give information and data indispensable to those for whom it is intended, thus surpassing in value any other book of its character; to economize the labor of the practical man, and to render his every-day calculations easy and comprehensive. It will be found to be one of the most complete and valuable practical books ever published.

Cabinetmaker's and Upholsterer's Companion,

Comprising the Rudiments and Principles of Cabinet-making and Upholstery, with Familiar Instructions, illustrated by Examples for attaining a proficiency in the Art of Drawing, as applicable to Cabinet Work; the processes of Veneering, Inlaying, and Buhl Work; the Art of Dyeing and Staining Wood, Bone, Tortoise Shell, etc. Directions for Lackering, Japanning, and Varnishing; to make French Polish; to prepare the best Glues, Cements, and Compositions, and a number of Receipts particularly useful for Workmen generally. By J. STOKES. In one volume, 12mo. With Illustrations,.......... $1.00

"A large amount of practical information, of great service to all concerned in those branches of business."—*Ohio State Journal.*

Campion. A Practical Treatise on Mechanical Engineering;

Comprising Metallurgy, Moulding, Casting, Forging Tools, Workshop Machinery, Mechanical Manipulation, Manufacture of Steam Engine, etc., etc. Illustrated with 28 plates of Boilers, Steam Engines, Workshop Machinery etc., and 91 Wood Engravings; with an Appendix on the Analysis of Iron and Iron Ores. By Francis Campin, C. E., President of the Civil and Mechanical Engineers' Society, etc. $7.00

Celnart. The Perfumer.

From the French of Madame Celnart; with additions by Professor H. Dussauce. 8vo. (*In press.*)

Colburn. The Locomotive Engine;

Including a Description of its Structure, Rules for Estimating its Capabilities, and Practical Observations on its Construction and Management. By ZERAH COLBURN. Illustrated. A new edition. 12mo, ... $1.00

"It is the most practical and generally useful work on the Steam Engine that we have seen."—*Boston Traveler.*"

Daguerreotypist and Photographer's Companion.

12mo., cloth, ... $1.00

Distiller (The Complete Practical).

By M. LAFAYETTE BYRN, M.D. With Illustrations. 12mo. $1.25

"So simplified. that it is adapted not only to the use of extensive Distillers, but for every farmer, or others who may want to engage in Distilling."—*Banner of the Union.*

Dussauce. Practical Treatise

ON THE FABRICATION OF MATCHES, GUN COTTON, AND FULMINATING POWDERS. By Prof. H. Dussauce. 12mo., ... $3.00

CONTENTS.—*Phosphorus.*—History of Phosphorus; Physical Properties; Chemical Properties; Natural State; Preparation of White Phosphorus; Amorphous Phosphorus, and Benoxide of Lead. *Matches.*—Preparation of Wooden Matches; Matches inflammable by rubbing, without noise; Common Lucifer Matches: Matches without Phosphorus; Candle Matches; Matches with Amorphous Phosphorus; Matches and Rubbers without Phosphorus. *Gun Cotton.*—Properties; Preparation; Paper Powder; use of Cotton and Paper Powders for Fulminating Primers, etc.; Preparation of Fulminating Primers, etc., etc.

Dussauce. Chemical Receipt Book:

A General Formulary for the Fabrication of Leading Chemicals, and their Application to the Arts, Manufactures, Metallurgy, and Agriculture. By Prof. H. Dussauce. (*In press.*)

DYEING, CALICO PRINTING, COLORS, COTTON SPINNING, AND WOOLEN MANUFACTURE.

Baird. The American Cotton Spinner, and Manager's and Carder's Guide:

A Practical Treatise on Cotton Spinning; giving the Dimensions and Speed of Machinery, Draught and Twist Calculations, etc.; with Notices of recent Improvements: together with Rules and Examples for making changes in the sizes and numbers of Roving and Yarn. Compiled from the papers of the late Robert H. Baird. 12mo...$1.25

Capron De Dole. Dussauce. Blues and Carmines of Indigo:

A Practical Treatise on the Fabrication of every Commercial Product derived from Indigo. By Felicien Capron de Dole. Translated, with important additions, by Professor H. Dussauce. 12mo..................................$2.50

Chemistry Applied to Dyeing.

By James Napier, F. C. S. Illustrated. 12mo........$2.00

CONTENTS.—*General Properties of Matter.*—Heat, Light, Elements of Matter, Chemical Affinity. *Non-Metallic Substances.*—Oxygen, Hydrogen, Nitrogen, Chlorine, Sulphur, Selenium, Phosphorus, Iodine, Bromine, Fluorine, Silicum, Boron, Carbon. *Metallic Substances.*—General Properties of Metals, Potassium, Sodium, Lithium, Soap, Barium, Strontium, Calcium, Magnesium, Alminum, Manganese, Iron, Cobalt, Nickel, Zinc, Cadmium, Copper, Lead, Bismuth, Tin, Titanium, Chromium, Vanadium, Tungstenum or Wolfram, Molybdenum, Tellarium, Arsenic, Antimony, Uranium, Cerium, Mercury, Silver, Gold, Platinum, Palladium, Iridium, Osmium, Rhodium, Lanthanium. *Mordants.*—Red Spirits, Barwood Spirits, Plumb Spirits, Yellow Spirits, Nitrate of Iron, Acetate of Alumina, Black Iron Liquor, Iron and Tin for Royal Blues, Acetate of Copper. *Vegetable Matters used in Dyeing.*—Galls, Sumach, Catechu, Indigo, Logwood, Brazil-woods, Sandal-wood, Barwood, Camwood, Fustic, Young Fustic, Bark or Quercitron, Flavine, Weld or Wold, Turmeric, Persian Berries, Safflower, Madder, Munjeet, Annota, Alkanet Root, Archil. *Proposed New Vegetable Dyes.*—Sooranjee, Carajuru, Wongshy, Aloes, Pittacal, Barbary Root. *Animal Matters used in Dyeing.*—Cochineal, Lake or Lac, Kerms.

This will be found one of the most valuable books on the subject of dyeing, ever published in this country.

Dussauce. Treatise on the Coloring Matters Derived from Coal Tar;

Their Practical Application in Dyeing Cotton, Wool, and

www.ingramcontent.com/pod-product-compliance
Lightning Source LLC
LaVergne TN
LVHW011231110826
845150LV00006B/1610